Anything that Rolls on Wheels

The People and Products of Montpelier Manufacturing

By Peter Costisick

Driveline
PUBLISHING

First published in 2005 by
Driveline Publishing
P.O. Box 838 • Yellow Springs, OH 45387

Book design by Suzanne Stevens

This book constitutes a non-commercial history of a commerical farm equipment manufacturer which has been published for educational purposes. All manufacturers' names, trademarks, model designations, examples of advertising and/or other photography are shown only for the purpose of identifying the related products and corporate entities and do not carry any implied product or policy endorsement.

All pictures, illustrations and photographs in this book are from the owner's private collections unless otherwise indicated.

Printed in the United States by Batson Printing

ISBN 0-9660751-6-1

. . . and the fire shall try every man's work of what sort it is . . . If any man's work shall be burned, he shall suffer loss: but he himself shall be saved: yet so as by fire.

I Corinthians 3: 13-15, KJV

Pete Costisick grew up in the western suburbs of Syracuse, New York. College was more affordable in the Midwest, so he chose to pursue a mechanical engineering degree at the Indiana Institute of Technology in Fort Wayne, graduating in 1970. He spent his entire engineering career in various product design and development positions at Cummins Engine Company, retiring in 2002. Previously published works in the area of automotive history include articles on Plymouth Barracudas and late 1950's DeSotos for the WPC News and a history of the Eisenhauer Twin Engine Freighter and X2 trucks, published by the American Truck Historical Society in Wheels of Time magazine. Currently, he is a contributing author for Vintage Truck magazine, researching and writing for the "Delivery Designs" series of articles. He's married to Pam, a terrific and very understanding wife and is blessed with Troy and Amy, two terrific children that have left the nest to pursue God's calling on their lives. He feels privileged to have hooked up with Driveline Publications and the talents of Pat Ertel and Suzanne Stevens in bringing this book to its readers.

Table of Contents

A 1964 Chevrolet 1½-ton chassis formed the basis for this large Side-Aisle Delivery truck operated by the Chicago Sun-Times newspaper.

A Montpelier Side-Aisle Delivery body mated to a shortened 1963 Ford F-350 chassis. These trucks were operated by a two-man crew consisting of a driver and a person in the cargo compartment who tossed bundled newspapers along the curb for subsequent delivery to individual homes. Montpelier Manufacturing enjoyed a steady business from large metropolitan newspapers, particularly those in Chicago and Philadelphia.

Hostess snack products were also a product of the Continental Baking Company in the early 1960's and Montpelier bodied truck were relied upon to deliver Hostess products from warehouses to local stores. The lightweight aluminum and fiberglass body construction improved vehicle acceleration and fuel economy. Lighter weight also reduced strain on the driveline that translated into lower maintenance costs and longer life prior to overhaul.

A 1964 high-top Urban model, the shorter of two wheelbase versions that the Montpelier Division produced for the Railway Express Agency on a Dodge 1½-ton chassis.

Foreward

Have you ever tried to chronicle the history of a company that was in business for thirty-eight years before ceasing operations nearly forty years ago? I hadn't either until I embarked on such an undertaking on December 23, 2003. I was aware of the existence of the Montpelier Manufacturing Company and thought that uncovering its history of truck body production would be an interesting pursuit but knew very little about the company. I stopped in at the Montpelier (Ohio) Public Library in the naïve expectation that a friendly, bespectacled, soft-spoken librarian would be able to drop everything I needed into a neat folder for me and I'd be on my way. Ridiculous! Instead, I was directed to a massive file cabinet filled with reels of microfilm. On a brief holiday visit two hundred miles from home, how was I going to be able to sift through nearly forty years of microfilmed newspapers to find randomly published articles that provided sporadic and superficial news about my subject of interest? I had previously and successfully undertaken much smaller ventures into the annals of automotive history and was fortunate enough to find everything I needed for a comprehensive story from no more than three authoritative sources of information. I had been spoiled by the simplicity of my previous exploits. Disenchantment set in soon after I popped the first microfilm cartridge into the viewer and started browsing.

Fortunately, I was rescued that morning by Al Benjamin. Al made one of his frequent visits to the library that morning and one of the librarians discreetly pulled him aside and said, "Al, you better help that hapless chap over there at the microfilm reader. From the way he's going about his business, I don't think that he could find his backside with both of his hands." Al is the premier local historian of Williams County, Ohio. He first directed me to a large volume that he co-authored with another Williams County native. A brief history of virtually every business enterprise that had ever existed in Montpelier and the surrounding area was contained within it. Discouragement began to wane slightly as I read Al's historical overview of Montpelier Manufacturing.

In the meantime, arrangements were made to meet with Carlton Gray who now lives in the stately East Main Street home where J. D. Hill, President of Montpelier Manufacturing until his death in 1939, once resided. Carlton loaned me two boxes of Montpelier Manufacturing artifacts, most of which he found in the attic when he purchased the home. Without hesitation, he took framed Montpelier Manufacturing documents off the walls of his home and included them in the stack of loaned items. Prospects for successfully researching and assembling the Montpelier Manufacturing story were improving. I had quite a bit of material to take home and absorb.

Al Benjamin...he's quite a guy and the incredible and sustained help he provided defies description. He made copies for me of all Montpelier Manufacturing news clippings he had in his possession. He then painstakingly searched those many reels of library microfilm that had stopped me dead in my tracks and found still more information. If answers to my questions weren't found in newspaper clippings, he would search elsewhere until the answers were found. He arranged interviews with over thirty former Montpelier Manufacturing employees or their relatives for me. There were times when I was unable to travel to Montpelier (or to Texas or Florida) for a face-to-face interview. In those situations, Al would deploy the skills and benefits acquired as a phone company retiree and initiate three-way long distance phone calls between the interviewee, himself and me and record the entire interview on tape for me. The project is now finished and the product of our work is in your hands. I'm no longer the naïve student of automotive history that I once was but, thanks to Al, I'm more spoiled than ever.

Is this book a perfect account of Montpelier Manufacturing history? I'd be the first to admit that it isn't. Yet the anecdotal stories gleaned from over thirty interviews and a thorough study of news clippings and surviving documents should be sufficient to give the reader

a good feel for the people and products of Montpelier Manufacturing. Is the historical account error free? No, that's simply not possible. Not having been an eyewitness to all of the events in the company's thirty-eight year history, I am no doubt guilty of at least a few errors, plenty of omissions and a certain degree of misinterpretation. Perhaps some of the people I interviewed embellished their personal history with the company a bit, too. I will admit to doing my very best to capture all that I could into a history that I hope is definitive enough to say that the job is done, once and for all.

Thanks to those of you who spurred me on with your enthusiastic support. I intend to keep an eye on you and will do all I can to encourage you in your exploits, just as you did me.

The following people were incredibly kind and hospitable to me during interviews and, in many cases, follow-up visits or phone calls. Each person either worked at Montpelier Manufacturing or is the relative of a former Montpelier Manufacturing employee.

Nan Bassett	Larry Bauer	Chuck Brigle
George Cornell	Kay Cox	Don Davis
Clare Dilworth	Ric Dye	Clarence Fisher
Bill Gipple	Al Goebel	Paul Goshorn
Carlton Gray	Frank Haase	Leland Haines
Wayne Henry	Paul Houk	Hubert Hutchinson
Jerry Kissinger	Bill Kunsman	Leo Mansfield
Denver Miller	Ed Miller	Keith Miller
Henry Pape	Martha Sapp	Harry A. Schwartz, Jr.
James Schwartz	Jack Shaffer	Lyle Shoup
Bonnie Stahl	Marcielle Starr	Roger Thorpe
Tamara Thorpe	Helen Wilson	

Thanks also to Al Benjamin, Clare Dilworth, Carlton Gray, Frank Haase, William Priest, Jr., James H. Schwartz and Harry A. Schwartz, Jr. for providing the vast majority of original photos and some of the few company documents that have survived over the past forty years or more. Since starting on this project, I was able to acquire a few more of these scarce items for a small collection of my own. The photos and graphics compiled for this book came everywhere from Belarus to British Columbia and from Michigan to Texas. Some of the photos and company documents that made it into this book barely survived their long hibernation and show signs of distress. They were deemed to be vital elements in presenting a complete story and were included, despite their flaws.

Bob Adler, a leading authority on Chevrolet light duty trucks on the 1940's and 1950's, was helpful in pinning down the model years for the Chevrolet cowl and chassis units that Montpelier used to promote its Side-Aisle Delivery bodies in various factory photos and sales sheets. Terri Hurd is a fellow church member and former co-worker. Also a computer aided drafting instructor, she did a masterful job producing a CAD drawing of the Montpelier Manufacturing factory layout and the surrounding property as it appeared in 1951.

The title of this book is derived from a direct quote by Harry Schwartz, the driving force behind Montpelier Manufacturing for over thirty-four years. Seeking military contracts from the War Department in the early phases of World War II, Harry was asked by an Army colonel what his company was capable of producing. ***"Anything That Rolls on Wheels"*** was Harry's straightforward and confident reply.

Our fast paced society has goaded us into being insistent on prompt answers and immediate results. We crave an answer or a certain product, i.e., the right hand side of the equation, and could care less about all of the terms and factors on the opposite side of the equal sign. The products of Montpelier Manufacturing were primarily truck bodies, mounted on chassis supplied by major light and medium duty truck producers. If that's all that interests you, then you'll be content to merely review the many photographs and other illustrations in this book. I would encourage you, however, ***not*** to neglect the many factors on the opposite side of the equation. How did the social and economic climate influence management and factory workers? Whose thumbprint was on the design of a particular truck body? What was the competitive environment and how well did Montpelier Manufacturing respond to it? Was management keen and progressive or passive and bound by old methods and traditions? How did management and workers respond to success or adversity? Whose blood and sweat went into each truck that rolled out of the factory's doors? What internal and external battles were won or lost? What was the ultimate achievement? How about the ultimate mistake? *What would you have done as a secretary, welder, foreman, shop superintendent or general manager?* Read the pages that follow and imagine yourself among ***The People and Products of Montpelier Manufacturing***. Enjoy mingling with them and determine what you would have done had you been in their midst.

I

Rising from the Ashes of Failure

The Montpelier Commercial Association (forerunner to the city's Chamber of Commerce) was, as any such association would be in the Roaring '20's, eager to bring industry to the community and thereby secure future prosperity for its citizens. The association's unbridled zeal is reflected in an August 1920 *Leader-Enterprise* newspaper ad that compelled (many would say coerced) local residents to, "... get in on the ground floor" by investing in the William A. Waggoner Talking Machine Company, hailed by the association as a new and sure-fire local business venture.

Human nature being as suspicious as it is today, a sales appeal of this type would probably have scam written all over it. However, a sufficient number of shares were sold to enable the new company to start erecting its manufacturing facility on Montpelier's Randolph Street, at the extreme southwest edge of town. Unfortunately, that was about all that the fledgling talking machine company accomplished prior to vanishing into thin air.

The debacle that was the Waggoner Talking Machine Company was quietly swept under the carpet by the Montpelier Commercial Association. Only seven months

Montpelier Manufacturing in 1932, five years after it set up operations in the former Turnbull Wagon Factory on Randolph Street.

later the association gushed with enthusiasm as it announced that "A new future for Montpelier seems assured through the efforts of the Commercial Association and the stockholders and directors of the William A. Waggoner Company." In discussions with H. L. Colt, a long-time director of the Turnbull Wagon Company, the Commercial Association learned that there was interest in renewing manufacture of Turnbull wagons. Colt was searching for a suitable facility with good shipping infrastructure that existed close to the wagon manufacturer's original plant in Defiance, Ohio. The Turnbull Wagon Company had ceased wagon manufacturing operations in Defiance in 1920 after 44 years of operation. Turnbull's Defiance plant and certain other company assets were combined to form the basis for a new manufacturing venture, the Defiance Motor Truck Company.

With a majority of stockholders in the now defunct William A. Waggoner Talking Machine Company approving the transfer of their Waggoner stock to the restructured Turnbull Wagon Company, plans were made to transfer all available Turnbull assets into the vacant Waggoner building.

Fully aware that wagon manufacture had been significantly curtailed due to the growing popularity of motor trucks, Mr. Colt and the new board of directors still felt that success for the revitalized Turnbull Wagon Company was achievable with skillful management. They perceived that a great majority of farmers were still committed to horse drawn wagons, something that they felt was disregarded by those companies that were moving into motor vehicle production. From an eternal optimist's perspective, the departure of several competitors from the wagon manufacturing business could only mean good news for the restructured Turnbull enterprise. A bigger slice of the wagon market was expected to come Turnbull's way as Studebaker of South Bend and Milburn of Toledo abandoned wagon production and converted to motor vehicle manufacture. The town of Montpelier eagerly welcomed the wagon business that the Defiance company cast aside. Mid-May found Mr. Colt and Turnbull's administrative staff settled into temporary office quarters on the third floor of the Masonic Temple as plans were coming together for transfer of the wagon tooling from Defiance to Montpelier.

In assessing the value of the business deal so cleverly engineered by the shrewd members of Montpelier's Commercial Association, Baker's *Leader-Enterprise* article encouraged readers to look beyond the mere physical assets of the company to the experienced workmen, the geographically distributed sales agencies and the highly esteemed Turnbull reputation as a further guarantee of the operation's success. From all points of view, all of the risks that were inherent in the Waggoner Talking Machine Company were conspicuously absent in the reorganized Turnbull Company. Baker's highly persuasive article easily established his credibility among Montpelier's citizens as Turnbull's eminently capable sales manager.

With full assurance that the wagon manufacturing business was coming to Montpelier, attention turned on May 18, 1921 to preparing the partially constructed building purchased from the receivers of the Waggoner Company for wagon production. Three days into the effort, the roof of the 100-foot wide by 300-foot long building was completed and efforts were then devoted to flooring. Heavy loads imposed on the floor during wagon manufacture made it necessary to install a floor that was four times as strong as the one initially planned by the Waggoner Company. Installation of floor joists and timbers followed pouring of additional concrete piers.

In February 1923 the Turnbull board of directors was elected for the ensuing year and the size of the board was increased from nine to fifteen members. Directors were gratified at the steadily increasing demand for wagons and business prospects for 1923 were bright.

However, favorable reports on the company's viability were conspicuously absent as 1923 wore on. The inevitable reality of advancements in surface transportation fell hard upon Montpelier's Turnbull Wagon Company in 1925 and the only recourse available to Mr. Colt and the board of directors was that of declaring bankruptcy. The obligatory public notice in the May 21, 1925 edition of Montpelier's *Leader-Enterprise* stated that book accounts and notes of the bankrupt company would be sold at public auction the following day, May 22. When the dust had settled, assets and liabilities of the Turnbull Wagon Company were $21,926 and $175,000 respectively.

Ownership of the property remained in the hands of a local group for only a short time. By November the building was sold to the A. J. Miller Company of Defiance whose short term intentions were to resume production of Turnbull wagons while installing equipment for the eventual production of various automobile and truck bodies destined for wholesale grocery, hearse, and ambulance markets. The Miller Company was no newcomer to the manufacture of motor vehicle bodies; they were the parent company for the Defiance Motor Truck Company and another body manufacturing plant in Bellefontaine, Ohio.

The A. J. Miller Company soon found that it had more than it could do to manage its Defiance and Bellefontaine plants and reneged on its plan to start wagon and truck body production in the former Turnbull factory. Ownership of the factory reverted to G. Grant Stahl, a prominent Montpelier businessman. Stahl had little difficulty in finding a buyer for the Turnbull plant, and it was a group whose aspirations were very similar to those articulated by the Miller Company in late 1925.

The new firm was in the process of being incorporated as the Montpelier Manufacturing Company as the sale of the Turnbull factory was finalized. The principal owners and members of the board of directors were Harry A. Schwartz of Cleveland and J. F. Robertson of Defiance, most recently employed at the Motor Coach Division of the Yellow Cab Company of Chicago. M. W. Legant of Dayton, owner of a valuable patent on a device used to form bodies for coaches, hearses, ambulances, and similar commercial vehicles was also a board member and served as the company's Vice President and Factory Manager. Schwartz was elected to the offices of President and Treasurer while Robertson became the firm's Secretary and General Manager. Although incorporated as the Montpelier Manufacturing Company, local citizens rarely referred to it as such, preferring rather to simply call Montpelier's newest business enterprise, "the body works."

Unlike the Defiance Motor Truck Company, Montpelier Manufacturing wisely chose not to produce complete commercial vehicles. Instead, they would specialize in the production of standard and custom-tailored bodies mounted on complete chassis supplied by other vehicle manufacturers that Schwartz, Robertson and Legant had previously worked with in their professional careers. Based on their collective experience, the men felt that they would be capable of operating the plant at full capacity once it was fully converted and upgraded. They also planned to manufacture wagons on a small scale and maintain a parts inventory to meet the service needs of existing Turnbull Wagon owners. The logic behind that decision rested in the fact that a mill room would be required to produce wood framing for hearse, ambulance and delivery truck bodies and excess capacity in the mill room could be utilized for producing a limited amount of wagons and replacement wagon parts.

One of the first men hired by the new owners was Bill Sprankell. Bill came to Montpelier Manufacturing after roughly six years of employment at A. J. Miller Company's two plants in Defiance and Bellefontaine, Ohio. Sprankell's experience ideally complemented the talents of Schwartz, Robertson and Legant and the firm started its body manufacturing enterprise by initially producing ambulance and hearse bodies. Sprankell served the new enterprise initially as its assistant plant superintendent and, a year and a half later, became the plant's full superintendent. The rest of Bill Sprankell's professional career was spent at Montpelier Manufacturing.

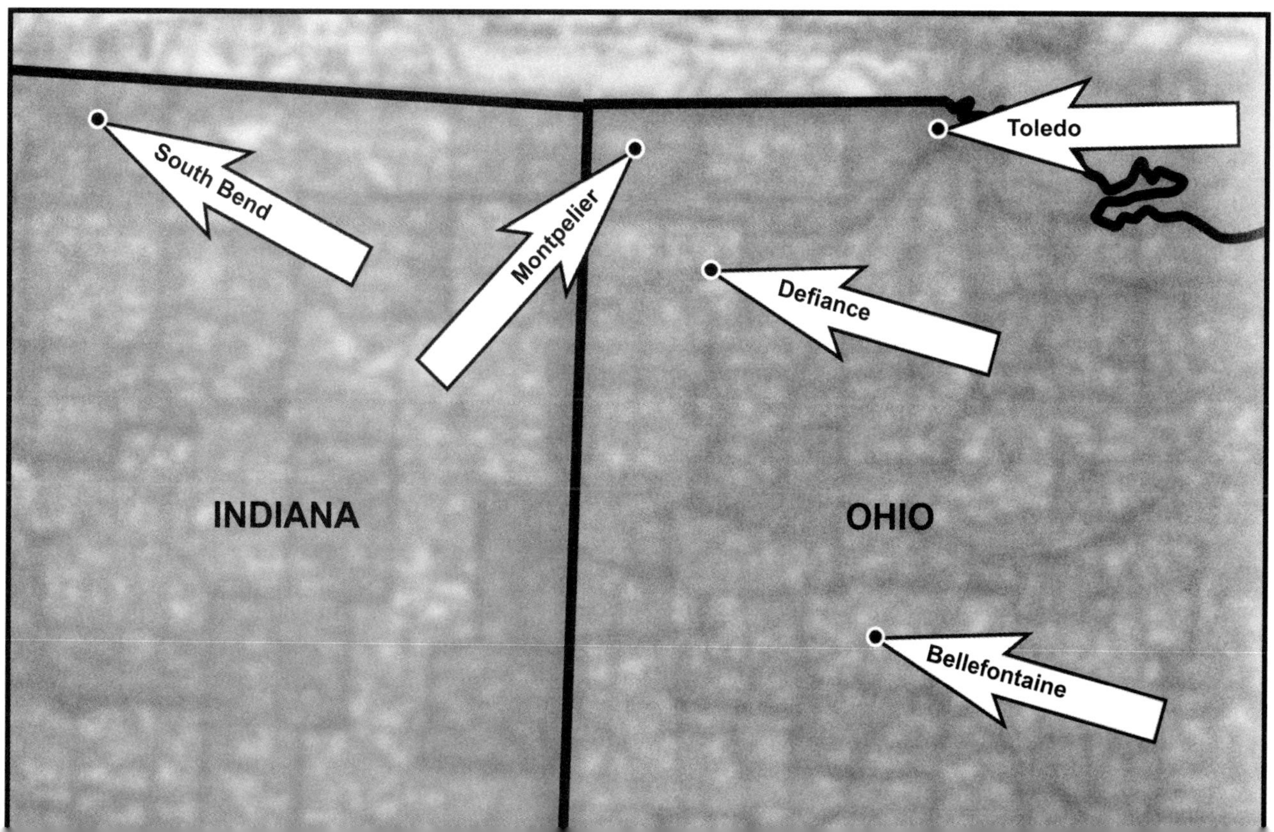

2

The Depression Years: Navigating Turbulent Waters

Montpelier Manufacturing Company was initially chartered in April 1927 at 1,000 shares. Schwartz owned 499 shares, Robertson 498 and Mary Robertson, T. T. Shaw, and Anna Sonnett each owned one share. Mary Robertson was the wife of J. F. Robertson, Shaw was the attorney who prepared Montpelier Manufacturing's incorporation papers, and Sonnett was Shaw's legal secretary.

On December 10, 1927, several stock transactions took place and shares in the company became equally divided, at 250 shares each, among Schwartz, Robertson, J. D. Hill and G. Grant Stahl. Hill and Stahl, two of Montpelier's most prominent businessmen, each paid $5,150 for their stake in the company. After a rapid succession of business startups and failures, it's rather amazing that the two prominent Montpelierites stepped

A Montpelier hearse body mounted on a 1930 Model A Ford cowl and chassis.

A Montpelier Manufacturing sedan delivery body mounted on a 1930 International cowl and chassis. Montpelier's deluxe sedan delivery bodies far surpassed the quality and appearance of the standard bodies offered by most truck manufacturers, The large, smooth radii that blend the top panel to the side and rear panels contrast sharply with box-like bodies that were still an industry standard.

forward to take a major stake in the new firm. Minutes of a board of directors meeting on that date also noted that M. W. Legant resigned as a director and vice president. One can only assume that he did so because the business was still in its infancy and wasn't generating enough revenue to pay him a salary on which he could survive. At the same board meeting, G. Grant Stahl was elected to fill the office of vice president that was vacated by Legant. At the January 23, 1928 board meeting, company officers and their titles were once again shuffled around. Directors unanimously elected J. D. Hill as President, Harry Schwartz as Vice President and General Manager, G. Grant Stahl as Vice President, Treasurer and Factory Manager and J. F. Robertson as Vice President, Secretary and Sales Manager.

After working through 1928 without meaningful compensation, Robertson followed Legant out the door, presumably to return to a more secure position in Toledo. He did, however, retain his stock in Montpelier Manufacturing and remained on the board of directors until 1940 when he sold his stake in the company to Harry Schwartz and Grace Hill. In the eleven years that followed his resignation as Secretary and General Manager, Robertson was a virtual nonentity as a director, only attending one board of directors meeting, a rather

An impressive Montpelier Manufacturing deluxe sedan delivery body mounted on a 1930 International cowl and chassis.

pathetic discharge of his responsibility to the company stockholders. Nevertheless, a majority of directors could still meet in spite of his consistent absence and he raised no objections to decisions made by the remaining directors, Hill, Schwartz and Stahl.

As owner of the Montpelier Creamery for the past 20 years, Grant Stahl had built the business into one of the most enterprising and envied business concerns in Northwest Ohio. In March 1927, Stahl abruptly sold the creamery to Page and Cox, a Detroit firm that was seeking an expanded territory and facilities for its egg and dairy business. At the time of the sale, Stahl indicated that he was undecided as to a future business venture.

Stahl also owned the Turnbull wagon property, having purchased it at auction after the bankrupt wagon company was dissolved. In April 1927, a few days before Montpelier Manufacturing was incorporated, Stahl sold the Turnbull property on contract to Schwartz and Robertson. Shortly thereafter, Montpelier Manufacturing was incorporated and the rights and responsibilities of the land contract were transferred from Schwartz and Robertson to the newly incorporated business. The fact that Stahl was a prominent and successful Montpelier businessman, owner of the contract on the Montpelier Manufacturing property, and a major investor in the business made him a logical choice to serve as a company director and vice president. Convinced by the collective experience of the Schwartz, Robertson, and Legant triumvirate and the quality of their business plan, he apparently decided to invest part of the proceeds from the sale of his creamery in the new commercial vehicle body building firm.

Under the terms of the $23,100 land contract, Montpelier Manufacturing agreed to make semi-annual interest payments to Stahl and to make clearly defined annual payments on the principal on or before each April 4th, from 1929 through 1934. Minutes of board of directors meetings from 1927 through May on 1939 contain no record of interest or principal payments on the land contract.

If Montpelier Manufacturing reneged on its land contract payments, at least it was gaining a tentative foothold in the truck body manufacturing business. The *Leader-Enterprise* of March 8, 1928 reported that "The plant is quite busy right now putting commercial bodies on Willys-Knight chassis, having 80 of these cars on its floors, 14 of which were received in one day. Besides this kind of work, the plant makes a specialty of equipment for undertakers."

Influenced at first by its past ties to the A. J. Miller Company, the firm started its body manufacturing enterprise by producing ambulance and hearse bodies. A photo of a Montpelier hearse body mounted to a 1930 Model A Ford chassis can be found on the first page of this chapter. Realizing, however, that the market for commercial delivery truck bodies had much greater potential for volume production, the company began focusing its efforts on delivery truck bodies built on 1/2-ton, 3/4 ton and 1-ton truck chassis, gradually phasing out of the ambulance and hearse body business in the 1930's.

It was apparent that the more exacting demands involved in the construction of high quality hearse bodies were mastered by Montpelier Manufacturing

A Montpelier body mated to a Willys cowl and chassis. Willys was a prime customer of Montpelier Manufacturing in it's earliest years.

and used to good advantage in the construction of their impressive sedan delivery bodies. Indeed, it may not be an exaggeration to say that the quality of Montpelier's sedan delivery bodies of the late 1920's and early 1930's was very nearly equal to some of the better custom coach built automobile bodies in the country. Montpelier bodies were distinguished by their excellent hardwood framing. All joints were mortised, glued, screwed and ironed. Top quality hardware was used for hinges and door latch and window crank mechanisms. Original factory photos of International and Willys-Knight cowl and chassis units equipped with Montpelier deluxe sedan delivery bodies can be found on the last two pages. The generous curves that formed the transition from the roof to the side and rear panels on Montpelier trucks bodies contrast sharply with the easier to fabricate squared-off joints on panel delivery bodies that were also available on International chassis of the same year. The photos on these pages were complemented by a Montpelier sales brochure dating back to 1930 that highlights "8 Features of Service and Comfort" and lists optional extra equipment for its full line of sedan type commercial bodies. A price list, on the next page, reveals initial prices for the full range of bodies and cabs offered by Montpelier Manufacturing in 1930. The handwritten notation on the price list indicates that prices were reduced by ten percent at some point in time, no doubt the result of the economic hardships brought on by the earliest phase of the Great Depression.

At the same time Montpelier was building made-to-order sedan delivery bodies, it was applying its design and manufacturing skills in the construction of truck cabs with integral sleeper compartments. The availability of their Model No. 105 De Luxe Sleeper Cab with its many distinguishing features was summarized in an illustrated single page specifications sheet published in August 1929. In contrast to the more rudimentary standard cabs found on 2-ton and larger

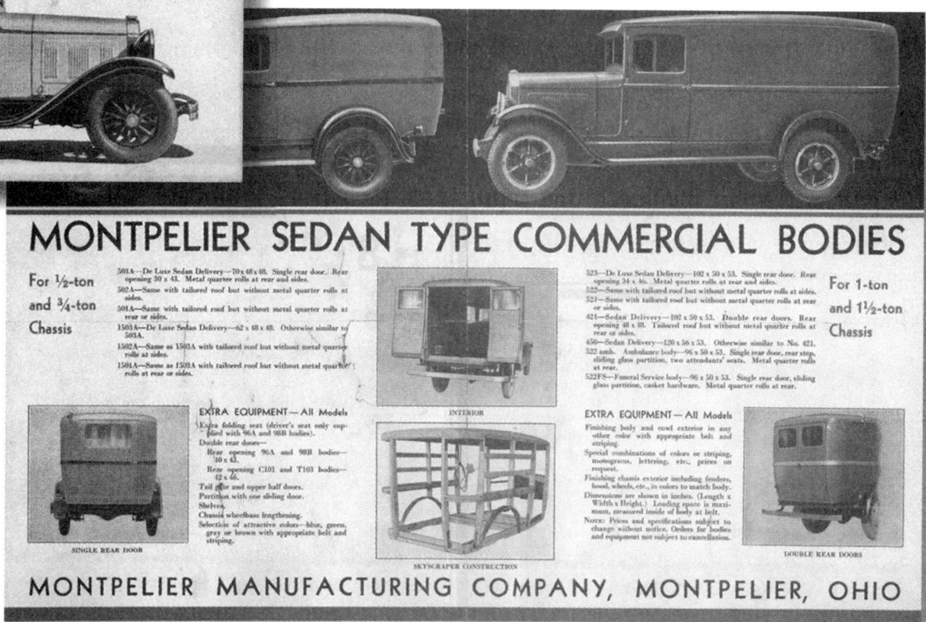

This circa 1930 Montpelier sales sheet points out the eight features that set a Montpelier sedan delivery body apart from the more commonly available standard offerings from light duty truck manufacturers. "Style, Dignity, Grace, Beauty and Ruggedness" were adjectives that Montpelier justifiably used to describe their commercial bodies.

Prices Reduced 10%

PRICE LIST
THE MONTPELIER MANUFACTURING COMPANY
Montpelier, Ohio
MONTPELIER BODIES AND CAB
(Effective June 20, 1930)

Model No.	*DESCRIPTION*	*List Price of Body Mounted on Chassis*
	For ½ Ton and ¾ Ton Chassis	
503A	DELUXE SEDAN DELIVERY—70x48x48. Single rear door. Rear opening 30x43. Metal quarter rolls at rear and sides.	$ 355.00
502A	Same with tailored roof but without metal quarter rolls at sides.	$ 325.00
501A	Same with tailored roof but without metal quarter rolls at rear or sides.	$ 295.00
1503A	DELUXE SEDAN DELIVERY—62x48x48. Otherwise similar to 503A.	$ 355.00
1502A	Same as 1503A with tailored roof but without metal quarter rolls at sides.	$ 325.00
1501A	Same as 1503A with tailored roof but without metal quarter rolls at rear or sides.	$ 295.00
	For 1 Ton and 1½ Ton Chassis	
523	DELUXE SEDAN DELIVERY—102x50x53. Single rear door. Rear opening 34x46. Metal quarter rolls at rear and sides.	$ 475.00
522	Same with tailored roof but without metal quarter rolls at sides.	$ 425.00
521	Same with tailored roof but without metal quarter rolls at rear or sides.	$ 395.00
421	SEDAN DELIVERY—102x50x53. Double rear doors Rear opening 48x48. Tailored roof but without metal quarter rolls at rear or sides.	$ 395.00
450	SEDAN DELIVERY—120x56x53. Otherwise similar to No. 421.	$ 495.00
522AMB	AMBULANCE BODY—96x50x53. Single rear door, rear step, sliding glass partition, two attendants' seats. Metal quarter rolls at rear.	$ 695.00
522FS	FUNERAL SERVICE BODY—96x50x53. Single rear door, sliding glass partition, casket hardware. Metal quarter rolls at rear.	$ 695.00
100	COUPE CAB (Blue) outside dimensions 30x56x54.	$ 135.00
	School Bus Bodies for 1 Ton and 1½ Ton Chassis	
39	Capacity 36 Children. 120x86x60.	$1395.00
40	Capacity 45 Children. 150x86x60.	$1525.00
41	Capacity 54 Children. 180x86x60.	$1765.00

Extra Equipment—All Models

Extra folding seat (driver's seat only supplied with ½ Ton and ¾ Ton Bodies.)	$ 12.00
DOUBLE REAR DOORS Rear opening ½ Ton and ¾ Ton bodies—40x43. Rear opening 1 Ton and 1½ Ton bodies—42x46.	$ 20.00
Tailgate and upper half doors	$ 25.00
Partition with one sliding door.	$ 35.00
Shelves, each	$ 15.00
Chassis wheelbase lengthening up to 36".	$ 140.00
Window in each side of 522AMB or 522FS set solid	$ 100.00
SELECTION OF ATTRACTIVE COLORS—Choice of Montpelier Standard *Blue, Green, Gray* or *Brown* with appropriate striping.	No extra cost.
Finishing body and cowl exterior in any other color with appropriate belt and striping.	$ 12.00
Special combinations of colors or striping, *Monograms, Lettering, etc.*	Prices on request
Finishing chassis exterior including fenders, hood, wheels, etc. in colors to match body.	$ 25.00
Re finishing wheels to match body	$ 5.00

NOTE: Body over all including belt, hood and cowl finished without extra charge, in choice of Montpelier Standard Blue, Green, Gray or Brown. Finish in other shades, color or colors combinations, see above for extra charge.

Dimensions are shown in inches. (Length x Width x Heigth.) Loading spaces are maximum, measured inside of body at belt.

NOTE: Prices and specifications subject to change without notice. Orders for bodies and equipment not subject to cancellation.

This 1930 Montpelier Manufacturing price list reveals the wide range of bodies the company offered. The handwritten note in the upper left corner reveals that prices were reduced at some date after the price list was published, presumably at attract more business in the face of a severely depressed economy.

trucks, the Montpelier specifications sheet announced that, "Passenger car practice of most approved type has been adapted to this impressive commercial cab." Montpelier sleeper cabs were mounted on flat-faced cowl and chassis assemblies that were diverted by the truck manufacturer to Montpelier's plant for installation of the cabs before the trucks were shipped to dealers for final delivery to the truck's new owner. The Montpelier specifications sheet reproduced on this page shows a Model No. 105 Cab installed on a late 1920's Mack Model BJ cowl and chassis unit.

Both exterior and interior cab proportions were quite generous. Length from the windshield to the back of the cab was 69 inches, height from the door sill to the cab top was 61 inches and cab width reached a maximum of 80 inches in the area of the sleeper compartment. The sleeping berth was 22 inches wide and 74 inches long. Generous radii blended all of the body panels into a rigid and very attractive unit in the same fashion as Montpelier's sedan delivery bodies, one of many distinguishing Montpelier features that competitive sleeper cabs lacked. Window glass was a full one-quarter inch thick in all locations. Rapid-action regulators raised and lowered the door windows with ease. The door windows were bound at the top edge with metal and slid effortlessly up and down in smooth metal-backed channels. Each generously radiused rear corner of the cab had a curved pane of glass at driver's eye level. A rare feature in its day, the extra window significantly improved the driver's peripheral visibility that otherwise would have been impeded by the integral sleeping berth.

It isn't known how much success Montpelier achieved with its made-to-order Model No. 105 Sleeper Cab or any similar cabs produced in the company's earliest years. Sleeper cabs are known to have appeared on trucks as far back as 1917 and truck manufacturers had at least a ten-year lead in establishing themselves as viable sources for trucks equipped with sleeper cabs of their own manufacture. The Model 105 and similar cabs produced by Montpelier were a highly desirable upgrade for those willing to pay for its excellent quality and many premium features not available on standard cabs offered by truck manufacturers.

A Montpelier Model 105 sleeper cab mounted on late 1920's Mack Model BJ cowl and chassis.

Montpelier

Passenger car practice of most approved type has been adapted to this impressive commercial cab.

Model No. 105

Montpelier DeLuxe Sleeper Cab

for Intercity Transportation, for Mounting on Tractor Chassis

Real riding and sleeping comfort, full visibility, good appearance, and an absolute maximum of rugged strength have been combined in this De Luxe Cab.

OUTSIDE DIMENSIONS—Length, front of windshield to back of cab, 69", height, sill to top, 61", width at widest point, 80".

FRAME—Seasoned hard wood free from sap, checks or knots. All joints mortised, glued, screwed and ironed.

BODY METAL—20 gauge auto body panel steel, formed to curves.

ROOF—Double texture deck material laid over twelve-ounce cotton, jute padded. 1¾" basswood slats mounted on crossbows and side rails. Metal drip moulding. Permanent visor construction.

DOORS—Sedan type, 32½"x50", on each side; mounted on heavy duty hinges; locks arranged with remote controls and lock handles; double rubber bumpers and heavy dovetail; rapid action regulator for windows which are of ¼" plate, metal bound at top edge and sliding in metal backed channels.

BODY WINDOWS—12" x 12" rounded glass in each rear corner, and 12" x 18" across rear; set solid in rubber channels.

SEATS AND BACKS—Individual section type; genuine leather, deep padded, over double deck springs.

BERTH—Two-section cushion with specially designed spring; genuine leather; heavily padded rail separating sleeping berth from back of seat. Berth dimensions 22" wide by 74" long.

INTERIOR—Artificial leather trim; large pockets; genuine walnut window trim; nickeled guard bars at rounded corner windows; bus type ventilator; dome light; rubber floor mat. Large storage or auxiliary gas tank space below berth. Compartment for clothing, etc., below driver's seat. Tool compartment with outside door at right rear of cab.

LIGHTS—Michigan markers at top front; dome light in ceiling; two plugs at outside rear for extra connections or extension trouble light.

WINDSHIELD—One-piece, ventilating type of ¼" plate glass.

FINISH—Duco—rubbed and polished.

STANDARD EQUIPMENT—Windshield wiper; rear vision mirror; scuff plates at doors; two green silk flags with nickel holders.

EXTRA EQUIPMENT—Auxiliary gas tank can be located under berth. Dash gasoline gauge. Dash cigar lighter. Trouble light, spot light, heater. Safety glass in windshield. Safety glass in doors. Berth curtains. Fire extinguisher. Auxiliary battery and tool boxes. Lettering. Monograms. Special chassis painting.

Curved panels not only give fine appea[illegible]e but greatest possi[illegible]e streng[illegible]

Montpelier c[illegible] are as rugged and sturdy as they look.

MONTPELIER MANUFACTURING COMPANY
MONTPELIER, OHIO

Form 105 2M—8-29 Mfrs. Adv., Inc., Toledo

MODEL 503A MOUNTED ON 1/2 OR 3/4 TON CHASSIS
LOADING SPACE AT BELT 72" BY 48" BY 50" HIGH

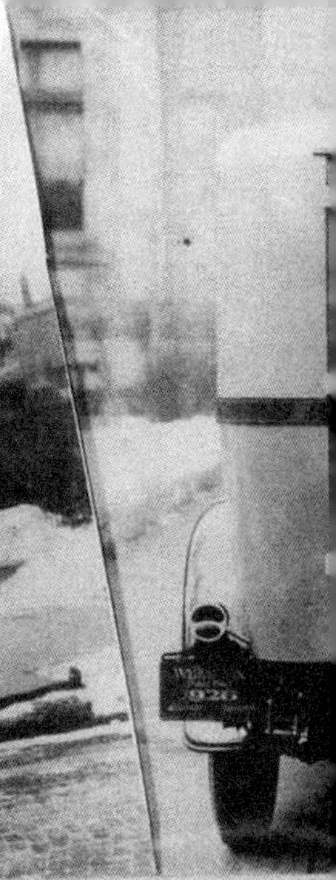

REA

THREE-QUARTER REAR VIEW MODEL 503A

M

D

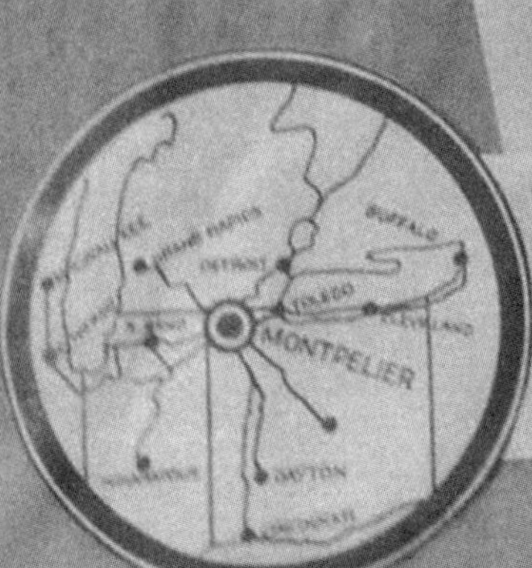

MONTPELIER M.

MONT

This simple sales sheet effectively portrays Montpelier Manufacturing's capabilities in its earliest days. It may have been included with a 1930 pricing sheet sent to dealers or prospective customers interested in sedan delivery trucks.

3
Catching the Next Wave with Side-Aisles

As a lower volume producer of finished bodies mounted to various truck chassis, Montpelier was able to individually tailor truck cabs and sedan delivery bodies to the needs of each truck buyer and his particular truck's application, something that the predominant light duty truck manufacturers were unable to do as effectively in their high volume assembly plants. In spite of its early success, Schwartz and his colleagues realized that Montpelier-bodied trucks still weren't substantially different from trucks offered by a host of truck manufacturers or competing body builders in the early 1930's. As a result, Montpelier Manufacturing strove to differentiate itself by offering truck bodies that were better suited to specific market segments than common sedan deliveries, panel trucks and cab and chassis trucks. A manufacturing strategy for capitalizing on the best available opportunities was gradually put in place. An important element in this strategy was Montpelier's pattern shop, which served as a bridge between the production floor and the engineering department which, in the 1930's, was Harry Schwartz.

Under the leadership of Bill Steinke, the pattern shop built initial prototypes of new or modified trucks or truck body options to verify that a new or revised design met all of its objectives before it went into full production. If called upon, the pattern shop would build

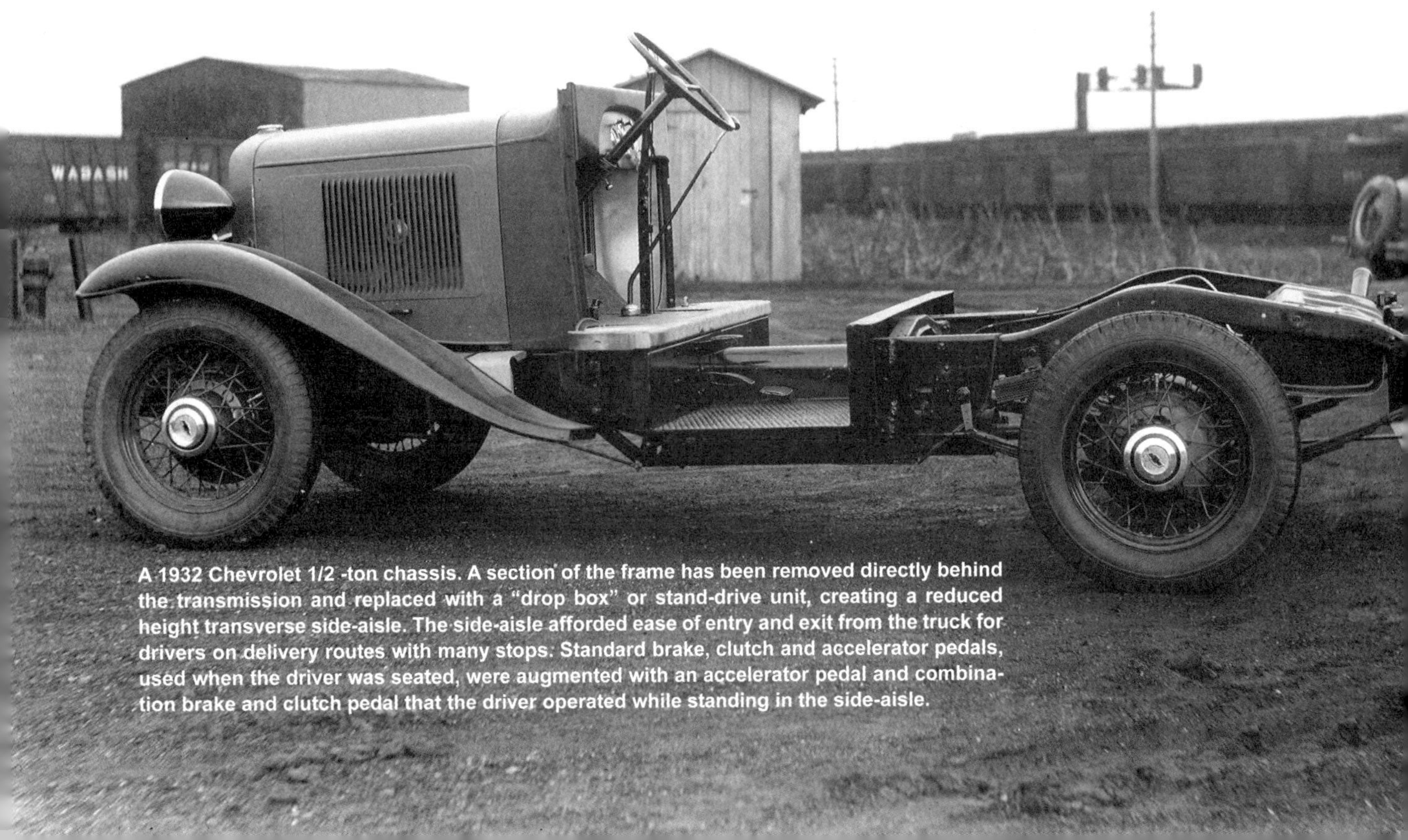

A 1932 Chevrolet 1/2 -ton chassis. A section of the frame has been removed directly behind the transmission and replaced with a "drop box" or stand-drive unit, creating a reduced height transverse side-aisle. The side-aisle afforded ease of entry and exit from the truck for drivers on delivery routes with many stops. Standard brake, clutch and accelerator pedals, used when the driver was seated, were augmented with an accelerator pedal and combination brake and clutch pedal that the driver operated while standing in the side-aisle.

an entirely new truck body or incorporate any changes or improvements (often customer specified) to an existing truck body, regardless of how large or small the changes were. Sometimes the changes were preceded by detailed engineering drawings but many times "reverse engineering" was more effective, i.e., on-the shop-floor trial-and-error design and development, followed lastly by engineering drawings.

Using past experience as a guide, each new or revised truck body panel started out as a flat pattern that was cut from 22 gage (.030 inch thick) or 24 gage (.025 inch thick) steel sheets with the aid of rotary or power shears. If the new pattern was destined to be a relatively simple part like a wheel housing, engine cover, or an interior shelf or storage rack, the sharp bends required for these panels were formed between the dies of a press brake with a minimum of difficulty. When a larger radius was called for in a body panel, patterns were inserted and clamped in Montpelier's metal rolling machine that operated on a principle that was akin to that of a conduit bender. The compound curves of fenders and the corner panels of steel tops were hammer formed on one or more of Montpelier's power hammers.

A trial fit of each new panel was performed in a nearby sub-assembly area as the metal forming phase neared completion. If one or more of the prototype panels weren't a hand-in-glove fit with its adjacent panels, it was reworked until a correct fit was achieved. If the rework failed to achieve the desired fit, the prototyping process started all over again with a new pattern that incorporated what was learned from the first trial pattern.

After a successful trial fit, the individual panels were welded and all seams between the panels were soldered. Pattern shop employees proceeded with the remaining phases of production for the prototype until the new or revised item was a finished product. The initial prototype was often presented to customers for their approval before an order was accepted and permanent production tooling and fixtures were developed. Once the design received customer acceptance, the pattern shop's involvement ended and responsibility for future production was transferred to the various metal fabricating, paint and final assembly departments on the shop floor.

Nearly all Montpelier employees directly involved in the manufacture of truck bodies or in modifying truck chassis supplied by U.S. light and medium duty truck manufacturers were compensated on a piece rate basis. Rather than being paid by the hour, each worker was paid a certain amount for each part or process he completed. Management favored this form of compensation because workers were paid for actual work output, not simply for the amount of time spent on the factory premises. From a management perspective, this kept labor costs in direct proportion to product output, enabling management to precisely determine at least one element of cost in the total cost of their finished product. On the surface, the piece rate compensation plan appeared to be equally favorable, perhaps even enticing, to shop employees with its inherent incentive to maximize earnings by being as productive as possible.

All was well in this seemingly equitable compensation system until management inevitably felt compelled to inject a certain degree of exploitation by placing a limit on the shop worker's daily earnings. If a worker improved his proficiency over time and required only six work hours instead of the previous eight to reach his daily compensation limit, it meant that he a.) sat idle or crawled into a hole and hid somewhere in the factory for two unproductive hours or he b.) continued to work and "banked" the completed work to turn in at some future date when business was slow and there wasn't enough work available to reach his daily pay limit. A third option, rarely if ever pursued, was to claim wages for work that exceeded the management limit on daily earnings. Pursuing this ill-advised option brought a few days of fleeting prosperity followed by a prompt management adjustment of the piece rate to a lower level of compensation for the completed work, thus bringing the worker back down to his daily earnings limit. Unwilling to take a proverbial "poke in the eye" as a reward for increasing productivity, shop workers gave no consideration whatsoever to improving productivity due to its attendant penalties.

Piece rate compensation was almost universal in the first three or four decades of the twentieth century. By the 1950's, however, all but a few diehard companies in the auto industry had abandoned it. Montpelier Manufacturing was inalterably committed to piece rate compensation and its effect on the prosperity of the business and its laborers will be pointed out from time to time in the pages that follow.

Side-Aisle, Drop-Frame, Stop-and-Go, House-to-House and Standdrive

Montpelier began to differentiate its products from the more generic trucks of the day by engineering its first Side-Aisle delivery bodies primarily for the delivery of either dairy or baked goods. These bodies were mounted on flat-faced cowl and chassis units for the most part supplied by Chevrolet. Trucks were offered on the customer's choice of either a 1/2-ton or 1½-ton Chevrolet truck chassis, modified by Montpelier as necessary to achieve the side-aisle feature. Montpelier collaborated with Chevrolet to market these trucks to dairies and bakeries starting in 1931. A sales folder published for the 1934 model year provides ample illustrations of the side-aisle, two sets of driver's controls, the well-established Chevrolet six cylinder truck engine and two vastly different cargo compartment designs, one tailored specifically for dairy deliveries and the other for delivering baked goods. The body could be ordered sans rear doors or with optional dual rear doors having a combined width of either 34 or 60 inches.

The side-aisle delivery truck derived its name from a lowered floor section across the full width of the truck at running board level, immediately behind the transmission and directly through the area where the driver was positioned. This transverse aisle gave the driver convenient curb-level side exits to his immediate right and left, hence the name side-aisle delivery. The side-aisle was interrupted only by a narrow steel housing that shielded the driver from the driveshaft or a torque tube.

The Step-N-Drive Corporation of Buffalo, New York first offered a patented form of the side-aisle delivery truck in the 1920's. Divco-Detroit was intent on offering delivery trucks with the same feature and acquired the patent rights by purchasing the Step-N-Drive Corporation in 1930. Both companies achieved

A Montpelier Manufacturing brochure pointed out many desirable features of the campany's Side-Aisle Delivery body mounted on a 1934 Chevrolet 1/2-ton cowl and chassis. This cover panel was complemented by seven additional fact-filled panels, some of which are seen on the following pages. Montpelier began installing these bodies on Chevrolet chassis i n 1931.

SPECIFICATIONS

Chevrolet-Montpelier Side-Aisle Delivery—half-ton and $1\frac{1}{2}$-ton models

CHASSIS:

Rated Capacity	$\frac{1}{2}$ Ton	$1\frac{1}{2}$ Ton
Wheelbase	112″	$113\frac{1}{8}$″

WEIGHT:

Chassis without body	2310 lbs.	3200 lbs.
Complete unit including body	3550 lbs.	4550 lbs.

CHASSIS DIMENSIONS:

Length overall	164″	171″
Width overall	$68\frac{13}{16}$″	73″
Face of dash to end of frame	$114\frac{1}{2}$″	$121\frac{1}{2}$″
Face of dash to center line of rear axle	$85\frac{1}{2}$″	$86\frac{5}{8}$″
Center line of rear axle to end of frame	29″	$34\frac{7}{8}$″
Face of dash to side aisle	$24\frac{1}{4}$″	$24\frac{1}{4}$″
Length of side aisle	$24\frac{1}{2}$″	$24\frac{1}{2}$″
Width of side aisle	$54\frac{1}{2}$″	$58\frac{1}{2}$″
Length back of side aisle to end of frame	$65\frac{3}{4}$″	$72\frac{3}{4}$″
Depth of side aisle to top of frame	8″	9″
Height from ground to top of frame at rear axle—loaded	$20\frac{11}{16}$″	$25\frac{3}{4}$″
Height from ground to floor of side aisle	$13\frac{3}{4}$″	$17\frac{1}{4}$″
Tread—Front wheels	$57\frac{9}{16}$″	$56\frac{13}{32}$″
Tread—Rear wheels	$57\frac{9}{16}$″	56″
Clearance—under front axle	$8\frac{3}{4}$″	$10\frac{1}{2}$″
Clearance—under side aisle	$8\frac{1}{4}$″	$8\frac{1}{4}$″
Clearance—under rear axle	$8\frac{3}{8}$″	$8\frac{3}{8}$″
Turning radius	21′ 10″	22′
Tires—Front	5.50—17″ 4 Ply	30 x 5″ 6 Ply
Tires—Rear	5.50—17″ 4 Ply	32 x 6″ 8 Ply
Rear axle reduction	4.11 to 1	6.17 to 1

CONTROLS:

Accelerator through toe board and in side aisle floor. Foot brake pedal through toe board. Foot brake pedal in side aisle. Combination clutch and brake pedal in side aisle. Master hand brake lever to right of driver to operate service brakes, and also emergency brake. Gear-shift lever.

Instrument board button for vacuum type automatic clutch control.

BODY DIMENSIONS:

Loading space length	64″	$76\frac{1}{2}$″
Loading space width	55″	60″
Loading space height	60″	60″
Side aisle height	74″	74″
Side door opening	26″ x $66\frac{1}{2}$″	26″ x $66\frac{1}{2}$″
Rear door opening	34″ or 60″ x 56″	34″ or 60″ x 56″
Capacity		56 milk cases

OVERALL DIMENSIONS, COMPLETE UNIT:

Length	170″	184″
Height—under load	$89\frac{1}{2}$″	$93\frac{1}{2}$″
Width	69″	73″
Height—loading space floor from ground at rear—under load	$27\frac{1}{2}$″	31″

The right is reserved to change specifications, colors or prices without incurring any responsibility with regard to trucks or chassis previously sold.

CHEVROLET-MONTPELIER SIDE-AISLE DELIVERY

Dairies, bakeries, and other firms with house-to-house routes need low-priced, large capacity delivery trucks—capable of the lowest possible operating cost. The Chevrolet-Montpelier Side-Aisle Delivery is expressly designed to meet these requirements.

As it can be operated from either a standing or sitting position, it is equally suitable for congested districts or less populous areas. The operator can reach any part of the loading space quickly and easily.

Chevrolet-Montpelier Side-Aisle Deliveries also provide the advantages of the Chevrolet truck chassis—six-cylinder valve-in-head engine, Blue-Flame combustion, heavy-duty rear axle, stabilized front end, and many other features that make Chevrolet the most economical truck you can buy. Any Chevrolet dealer is equipped to provide service at typically low Chevrolet rates.

Side-Aisle Deliveries are available on the Chevrolet 1½-ton chassis illustrated above, and also on the Chevrolet half-ton chassis. Body interiors for various types of service are described on the other side of this folder.

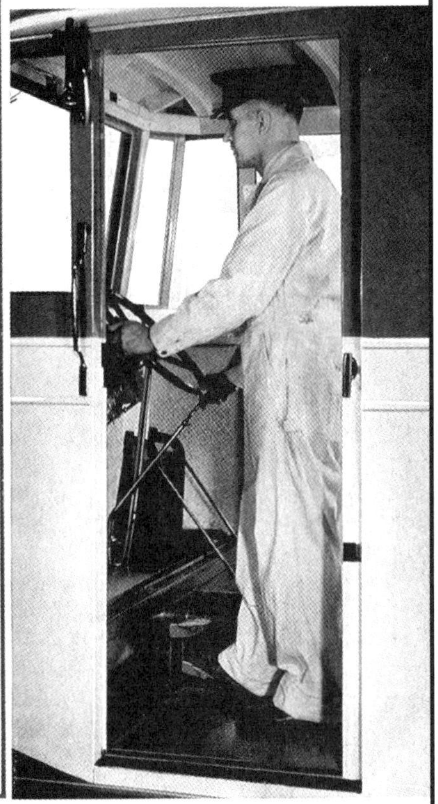

Montpelier Side-Aisle delivery bodies were designed for productivity and driver-friendliness in stop-and-go delivery applications like few other trucks (or horse-drawn wagons) before them.

the side-aisle feature by offering continuous frame rails with each siderail having a lowered section that started directly behind the transmission. Montpelier took a different approach to incorporating a side-aisle and apparently the differences were sufficient to avoid infringement on the Divco patent.

Jack Shaffer began working at Montpelier Manufacturing in 1954, two years after graduating from high school. He spent the majority of his time working for Wayne Bordner in the chassis department, modifying chassis for side-aisle delivery trucks in a manner that had remained consistent over the previous twenty years or more. Chassis modifications were performed under the shelter of a lean-to during months when the weather was mild. When temperatures dropped, the work was moved indoors to a garage area on the factory's north side. Jack recalls that roughly four chassis were prepped each day. Truck chassis with straight frame rails chassis received from a variety of light duty truck manufacturers were shortened or lengthened as needed for each particular truck build. Typically, twenty-five inches were removed from each frame rail, directly behind the transmission. A prefabricated and heavily reinforced steel stand-drive unit or "drop box" replaced the frame rail section. The drop box was formed from 10 gage (.135 inch thickness) and 4 gage (.234 inch thickness) steel and was every bit as sturdy as the original frame rail sections that it replaced.

Jigs were used to maintain frame alignment during welding. Fishplates were welded and bolted to the modified areas of the frame for reinforcement. Welds or areas where the chassis was ground would be brush painted chassis black to match the rest of the chassis. Driveshafts were taken to Montpelier's tool room where Jay Ransom altered their length and rebalanced them before the chassis department reinstalled them. Special rear bumpers were made and installed in the chassis department, some with a drop step that facilitated entry through the rear doors of the delivery body.

Chassis, when fully prepped, would go back into the company's north lot. When a modified chassis was needed for mounting a finished delivery body, it was rolled around the east side of the plant and through an overhead door on the plant's south side. A photo of a 1932 Chevrolet cowl and chassis with its drop box welded in place can be seen on the first page of this chapter.

The driver had the option of operating a side-aisle delivery truck in the normal seated position, using the standard hand and foot operated controls that came with the chassis, or he could move the pivoting driver's seat out of the side-aisle and operate the truck in a standing position. When standing, the driver used the standard steering wheel and floor mounted gearshift along with a separate set of foot-operated controls for the accelerator, clutch and brake protruding through the floor or the vertical panel of the side-aisle. Entering or exiting the truck via the side-aisle and full-length doorways on either side of the truck was almost as easy as walking through an open doorway in one's home. There were no steps to negotiate and no contortions were required to get into or out of the driving position, advantages that were greatly appreciated by route drivers who typically had their hands full of their wares as they entered or exited the truck. Sliding doors or optional "jackknife" folding doors at either end of the side-aisle could be left open or closed at the driver's discretion while the truck was in transit.

A New Set of Competitive Challenges

Though it was a well-conceived body design, the Side-Aisle delivery body was definitely not the exclusive domain of Montpelier Manufacturing. By 1932, truck manufacturers, including Divco-Detroit,

Sales literature emphasized Montpelier Manufacturing's ability to offer delivery bodies of uncompromised value, achieved by tailoring truck bodies to specific delivery applications.

LOADING SPACE—MILK DELIVERY

For milk delivery, the 1½-ton Chevrolet-Montpelier Side-Aisle Delivery is paneled on the sides and the floor with galvanized sheet steel. Rub-rails are provided on each side for easy sliding of milk cases. Galvanized steel racks are installed over each wheel house to support milk cases.

LOADING SPACE—BAKERY DELIVERY

For bakery service, the half-ton Chevrolet-Montpelier Side-Aisle Delivery is paneled on the sides with three-ply fir in natural wood finish. A second floor of matched wood is laid over waterproofed material to protect the flavor and freshness of baked goods from moisture, fumes and dirt.

Dodge, Ford, International, Kenworth, Stutz, Thorne, Twin Coach, Walker, Ward and White were already producing comparable trucks. In addition, companies besides Montpelier were busily engaged in mounting side-aisle bodies to flat-faced cowl chassis supplied by truck manufacturers. These included the DeKalb Wagon Company of DeKalb, Illinois, Boyertown Auto Body Works of Boyertown, Pennsylvania, Swift Body & Equipment Company of Saginaw, Michigan, the Standrive Company of Los Angeles, California, Metropolitan Body Company of Bridgeport Connecticut, The Edwards Iron Works, Inc. of South Bend, Indiana, McCabe-Powers of St. Louis, Missouri, and the Proctor-Keefe Body Company in Detroit, Michigan. The terminology used to describe this type of truck in the years following its emergence was just as varied as the names and geographic locations of their manufacturers. Drop-frame, stop-and-go, house-to-house and stand-drive trucks of the late 1920's and early 1930's were the essential equivalent to a Montpelier Side-Aisle.

Federal, Indiana, Reo and Studebaker entered the fray in 1932 and 1933. Truck body builders Hackney Brothers, the York-Hoover Body Corporation and Columbia Body and Equipment Co. were some of the more prominent truck body builders that entered the delivery truck market with similar bodies a few years later. Hackney built bodies of this type for nearly fifty years and, although their operations were sold and merged with another body builder in the mid-1990's, a Hackney & Sons insulated truck body is still available today.

The Stutz entry was the Pak-Age-Car that was first introduced by the Package Car Corporation of Chicago in 1926. Stutz started marketing this truck for the Package Car Corporation in 1927, taking over production for the struggling company in late 1931 and reengineering the truck, bringing the body and drivetrain up to contemporary standards. Although it was small and underpowered throughout most of its sixteen-year production run, the Pak-Age-Car was, at least in a few respects, decades ahead of its time.

Whether they realized it or not, Montpelier Manufacturing had entered a crowded field of truck manufacturers and body builders who offered trucks that competed directly with their Side-Aisle Delivery. However, the market was ripe for such vehicles and Montpelier's strategic partnership with Chevrolet gave the firm a distinct advantage. Radio and print media advertising exposed both rural and metropolitan families to "must have" durable goods and consumables. "Going to market" via one's personalized transportation was fraught with many problems and still not an option for everyone. Virtually any business owner providing a product or service was more than happy to deliver his goods for his customers and the side-aisle delivery truck was becoming the vehicle of choice for many of them. All of the manufacturers of trucks similar to Montpelier's Side-Aisle Delivery could take comfort in the fact that at least one competitor was steadily relinquishing market share to them, that being the delivery vehicle of the horse-drawn variety.

Montpelier Manufacturing was poised to take full advantage of Chevrolet's immediate name recognition, its extensive sales and service network and a highly tooled and very cost competitive chassis. For more than three decades, Montpelier Manufacturing produced thousands of side-aisle delivery bodies, eventually mounting them on chassis supplied by every U.S. light-duty truck manufacturer.

Even though Montpelier's Side-Aisle delivery bodies were on a solid footing in terms of innovation and customer satisfaction, the financial situation of the company was still very tenuous. From April 1936 through November 1938, Montpelier Manufacturing's board directed company officers to proceed with six bank loans totaling $27,000 and a seventh bank loan of up to an additional $10,000. Schwartz, Stahl and Hill also forfeited a combined $7,500 in 1936 salary to meet the company's immediate needs for available cash. In order for the company to secure the bank loans, Stahl and Hill provided each bank with a notarized agreement that "all of the indebtedness of said company to you shall be a prior claim over any indebtedness of said company due to us or either of us." Disregarding overdue payments on the 1927 land contract, Montpelier Manufacturing's indebtedness to Hill and Stahl as of October 5, 1938 stood at $53,382.19, $10,000 of which was owed to Stahl and the balance, consisting of two demand notes dated December 31, 1935, owed to Hill. Still confident in their ability to make the company's finances as viable as their delivery bodies, all who had a financial stake in the company encouraged Schwartz to forge ahead with new and more diversified products.

4

COE's and Urbans

COE Cabs

With steady drivetrain and chassis advancements, truck design in the 1930's underwent a gradual transformation with the aim of increasing cargo volume and payload capacity for a given overall vehicle length. Other demands placed on truck manufacturers by truck fleet owners and their drivers included improved maneuverability in traffic, smaller turning radii, improved forward and peripheral driver visibility and more uniform weight distribution between front and rear axles.

Dodge produced this brochure to introduce its customers to its first line of cab-over-engine trucks in 1937. Chassis conversion to forward control steering and body installation were performed and warranted separately by Montpelier Manufacturing.

An elaborate four-page Montpelier Manufacturing brochure promoting the company's cab-over-engine cab for 1938 Chevrolet trucks. The cabs were installed on Chevrolet 1½-ton chassis with wheelbases ranging from 108 to 157 inches. Montpelier COE cabs were phased out the following year when Chevrolet started producing their own COE truck cabs.

The most far-reaching response on the part of truck manufacturers to this set of requirements was the emergence of alternate truck cab designs, starting in the mid-1930's. While the driveline layout remained much the same, truck cabs were moved forward in varying degrees, making a larger percentage of the chassis behind the cab available for payload. Terms used to describe these alternate cab locations included snub-nose conventional, cab-over-engine (COE) and cab-forward. Montpelier entered the market for cabs falling into one or more of these categories starting in late 1936.

A lengthy front page article appearing in the February 25, 1937 *Leader-Enterprise* reported that, "According to Manager H. A. Schwartz, business for 1937 will exceed that of 1936, unless some untoward incident crops out. The Cab-Over-Engine which this firm has been producing has become so popular that other firms have entered the field" and "Orders for Cab-Over-Engine have been placed through the manufacturers of automobiles and the local firm has co-operated with them in creating this market, having received orders from International, General Motors, Dodge and others."

Montpelier's work with Chevrolet and Dodge as a supplier of COE cabs is well documented but no details outlining the company's collaboration with International have come to light. Chevrolet and Dodge both specified Montpelier COE cabs mounted on 1½-ton forward control truck chassis for the 1937 model year. Installation of the cabs, including conversion to forward control steering and any other chassis modifications required in order to install the cabs were performed by Montpelier. Montpelier's COE cabs for both Dodge and Chevrolet were almost indistinguishable from each other, the only visual differences between the two being badging.

Dodge chose to offer only the day cab version of Montpelier's COE cab for 1937, while Chevrolet also offered the sleeper cab version. One of Dodge's sales brochures for the COE truck lineup for 1937 appears on the first page of this chapter. Chevrolet's *Silver Book*, an annual publication containing data for all special equipment available on Chevrolet trucks,

provided Chevy truck dealers and prospective buyers with illustrations and a full set of specifications on all Montpelier-bodied Chevrolet trucks for 1937.

Montpelier developed a new day cab for Chevrolet for the 1938 model year. It was a much cleaner and more contemporary design that looked much like the highly tooled cabs Chevrolet produced for its conventional line of trucks. The new cab featured all steel exterior panels, a marked improvement over the 1937 cab which bore a heavy cloth decking on the roof, supported underneath by wadding and three-ply fir paneling. A sleeper cab version was not offered for 1938, presumably because of low demand and the high tooling costs that would have been associated with an integral sleeper version of the new 1938 cab. Montpelier's 1938 sales folder for Chevrolet trucks equipped with the handsome Montpelier COE cab can be seen on these two pages.

Chevrolet introduced its own COE cab for the 1939 model year. This design bore such had close resemblance to the COE cab that Montpelier offered in 1938 that some of the folks at Montpelier believed Chevrolet had appropriated the Montpelier design.

Dodge retained the somewhat unorthodox looking Montpelier COE cab from 1937 through 1939. By 1939, the generic grille design initially shared with Chevrolet was abandoned in favor of bright trim that so closely resembled the rest of Dodge's 1939 truck

A 1939 sales brochure promoted Montpelier's COE cab mounted on a Dodge 2-ton truck chassis. Chassis modifications required to accept the COE cab were performed and warranted by Montpelier. Montpelier COE cabs were also fitted to Dodge 1½-ton truck chassis. Dodge first started using these cabs in 1937, eventually replacing them with COE cabs manufactured in their own plants in 1940.

lineup. Dodge made expanded use of Montpelier COE cabs in 1939, offering them in its 2-ton TH Model chassis as well as the 1½-ton lineup. A 2-ton truck equipped with a Montpelier COE cab and a box van body appeared in a Dodge sales brochure covering its 2-ton and 3-ton trucks for 1939 and is reproduced on the next two pages.

Montpelierites who read the *Leader-Enterprise* news article might have concluded that Montpelier Manufacturing was at the forefront in the development of this new crop of COE truck cabs but in reality, that was not the case. Between 1934 and 1939, virtually all manufacturers of 2-ton and larger trucks began offering COE cabs, either manufactured in-house or purchased from other body builders. International trucks with COE cabs made their first appearance in limited numbers in 1934, the cabs being supplied by the Hendrickson Company of Lyons, Illinois. GMC introduced a COE truck in 1934. By 1936, International was in production with its C-300 and CS-300 trucks with COE cabs of their own manufacture. Studebaker and Reo were offering COE cabs in late 1935 while, at the same time, Ford offered the "Dearborn Line" of COE cabs manufactured by the Transportation Engineers Corporation for their trucks.

A 1939 International Metro forward control delivery truck. The Metropolitan Body Company of Bridgeport, Connecticut became a wholly owned subsidiary of International-Harvester in 1938 and began producing Metro delivery truck bodies exclusively for International beginning in the 1939 model year.

By rearranging machinery on the Montpelier Manufacturing production floor, an additional 6,000 square feet of floor space was gained, increasing total manufacturing floor space to 40,000 square feet. There were times, however, when twice as much floor space could have been utilized. New machinery purchased by the company was reluctantly stored outdoors under tent coverings until it could be integrated into the overall layout on the production floor. Axles, fenders, radiators and other components removed from incoming chassis accumulated in piles outside the plant and, for the sake of expedience, were sometimes sold at considerable loss.

In less than ten years, Montpelier Manufacturing had become the town's largest industry with an annual payroll that exceeded $100,000. In the fall of 1936, the city installed a new water main leading up to the plant, giving it an ample supply of water in the event of a fire. Realizing that it would take the fire department considerable time to reach the plant at the southwest edge of the city if a fire were to occur, company management expressed its intention to install a sprinkler system in the plant as soon as possible to further protect the plant from possible fire damage.

The COE Urban Van

The February 25, 1937 *Leader Enterprise* story went on to report that "the local plant is producing a new light delivery outfit which, it is expected, will create even a greater demand than did the *(Side-Aisle)* trucks. A photograph of the first unit of this new product was shown to one firm in Chicago recently and five of them were immediately ordered. They provide thirty to forty percent more space than the old style delivery outfits. There seems to be no question that the new product will prove to be extremely popular and that the demand will grow rapidly.

"In making these new delivery outfits the company has installed a great quantity of expensive machinery and will continue to add to their facilities of production. In many cases they almost completely rebuild the chassis in order to conform to the requirements of the body and cab to be installed. And although starting

from 'scratch,' the company is now in a position to handle any sort of work growing out of body building and chassis changing. Their product has attracted much attention and favorable comment wherever displayed and on the highways."

Although historical accounts of the late 1950's indicate that Montpelier's first prototype forward control delivery van was built on an International chassis in 1937, no verifiable record of such work is known to exist today.

News about Montpelier's COE cab and Cab Over Engine Urban Type Deluxe Van was contained in a single news release because the two new products were developed concurrently and shared many common components. Grilles, cowling, front fenders, front-hinged doors, windshield framing... basically the complete front clip of COE cabs and the largest Montpelier Urban Vans were identical from the doors forward. Areas where the two designs departed from one another consisted of roof panels and sheet metal directly aft of the side doors. Smaller Urban Vans shared the same cowling as the COE cabs and larger vans while offering sliding doors and larger windshield and front corner windows, details that were better suited to the frequent stop-and-go duty cycle anticipated for shorter wheelbase Urban Vans.

The new COE Urban Van offered a thirty to forty percent increase in cargo volume over the Side Aisle delivery trucks without a corresponding increase in overall vehicle dimensions. The more efficient configuration of the Urban Van moved the nominal center of the truck's payload considerably forward in comparison to a side-aisle truck, resulting in more uniform weight distribution between the front and rear axles. When mounted on a Chevrolet chassis the increased loading of the front axle meant that the standard Chevrolet 1½-ton front axle and springs had to be discarded in favor of heavier units.

No truck chassis would accommodate a Montpelier COE cab or Urban van body without first converting the steering box and linkage to forward control operation. The design and manufacturing tasks involved in the conversion were made a tad more difficult and expensive because the standard steering box had to be replaced with a heavier unit that was compatible with the increased weight over the front axle.

Montpelier and Chevrolet jointly promoted the new Montpelier Cab Over Engine Urban Type Deluxe Van in 1937 but by 1938 Chevrolet offered its 1½-ton chassis to several additional manufacturers of forward control delivery truck bodies, most notably Boyertown Auto Body Works of Boyertown, Pennsylvania and the

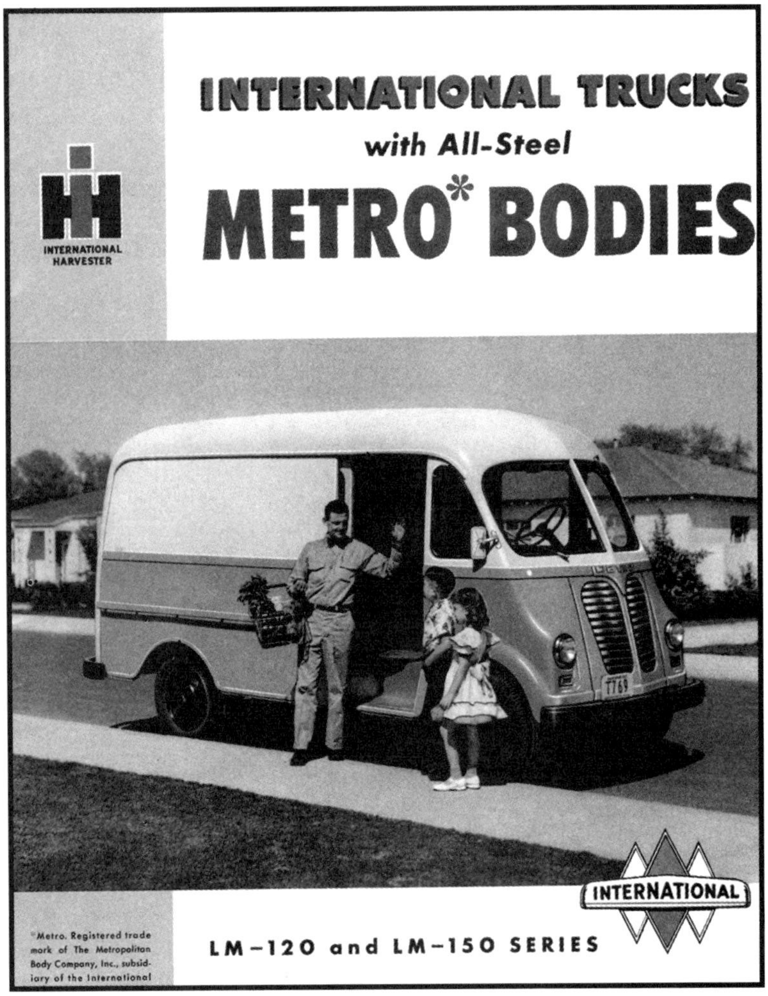

The rounded body design of International Metro delivery truck changed very little from its initial introduction in 1939 until 1966 when it was finally phased out and replaced by a more contemporary, squared-off design. The model shown on the cover of this International Metro sales brochure dates back to 1950. International's Metro series of forward control delivery trucks was one of Montpelier's more formidable competitors. Montpelier had the advantage of being able to tailor its truck bodies to the exacting needs of individual customers in a way that a higher volume producer like International was not as well equipped to do.

This ad promoting Dodge forward control delivery trucks for 1941 appeared in a Grand Rapids, Michigan newspaper. The testimonial of a local dairy business that operated these trucks was used to point out the truck's fine attributes to local dairies, bakeries, laundries and department stores. No mention was made of a body builder but many unique characteristics of the Montpelier Urban Panel delivery body are unmistakable.

Metropolitan Body Company. In 1939 the Metropolitan Body Company became a wholly owned subsidiary of International Harvester and ceased production of truck bodies for all of International's competitors. With International's strategic move to acquire a delivery truck body builder, one of Montpelier's more formidable competitors in the decades to come would no longer be the Metropolitan Body Company but its parent company, International-Harvester, and its very popular line of Metro delivery trucks. The earliest of the International Metros and a 1950 descendent are illustrated on these pages.

Dodge called its version of the first generation Urban Van the Job-Rated Montpelier Urban Panel Delivery and offered these trucks in cargo compartment lengths of either nine or eleven feet through the 1942 model year. Promotion of the 1941 models via the printed media took several forms, three of which are illustrated on the next page.

A page taken from a 1941 Dodge full line truck brochure shows the Montpelier Model 9 Urban Panel Delivery body installed on a 3/4-ton forward control truck chassis. A larger Model 11 body was also offered on Dodge's 1-ton chassis.

As was the case with Chevrolet, Montpelier wasn't the exclusive builder of delivery truck bodies for Dodge. At least one company, again the Metropolitan Body Company, is known to have built delivery truck bodies mounted to forward control Dodge chassis through 1938.

As the 1930's drew to a close, forward control delivery trucks of contemporary design were popping up like wild onions in the early spring. Truck manufacturers like International, Mack, Reo, Divco, Chevrolet, GMC, Stutz and White each had a forward control delivery truck of their own design on the market. Body builders that were just as ready and eager as Montpelier to install delivery bodies on any truck manufacturer's forward control truck chassis included Boyertown, Metropolitan, Gerstenslager, General and Gillig.

Dodge produced this highly detailed four-page sales brochure for Montpelier's 1941 Urban Panel Delivery bodies mounted on Dodge 3/4-ton and 1-ton chassis. Dodge's marketing effort on behalf of Montpelier's Urban was the most aggressive to date by a major light-duty truck manufacturer.

5
Competing for the Delivery Truck Market

With each new product that Montpelier Manufacturing released came hopes for an invigorated and extended lease on life for the company. Unfortunately, those hopes were never fully realized due to the less than exclusive nature of Montpelier's new products, the long list of competing body builders striving for the same business and the move on the part of some truck manufacturers to produce delivery truck bodies mounted to their own chassis within their own assembly plants.

MONTPELIER URBAN DELIVERY

For House-to-House Deliveries

Dealers selling the Montpelier Urban Delivery have found themselves in an advantageous competitive position for equipment of this type mounted on a suitable chassis, depending upon the type of service or load to be carried. The Montpelier Urban Delivery is ideal for Department Store, Bakery, Dry Cleaning, and other House-to-House Delivery Services. These retail delivery services have found the Montpelier Urban Forward Control Panel more effective, convenient and economical in cities where congested traffic conditions prevail, as well as in more remote localities. Almost EVERY dealer territory has at least one prospect for the Urban Forward Control Panel. A Montpelier sale means a three-way profit: (1) new chassis, (2) Montpelier Unit, and (3) optional equipment. Write, wire or telephone Montpelier for further information and prices.

The above view shows a Montpelier Urban Panel, Forward Control type, Model No. 9, 9 Ft. size, mounted on a suitable chassis of a capacity depending on the load to be carried. Note the sliding type side doors with running board level extending into the driver's compartment as far as the chassis frame will permit.

OPTIONAL EQUIPMENT:

Optional equipment at nominal cost, to suit customers' particular needs is available. Usual optional features are double rear doors over rear opening 51" wide by 58" high; partition separating driver's compartment from loading compartment with one or two sliding doors; steel rub rails at each body side; either one-or two-tone color; insulation; etc.

SPECIAL FEATURES:

Loading space dimensions for the Model No. 9 Urban Delivery are 9 Ft. long by 68" wide by 68" high at the center, suitable for installation on chassis with wheelbases from 108" to 120".

The Montpelier Urban Delivery, Model No. 11, has loading space dimensions 11 Ft. long by 68" wide by 68" high at the highest point in the center, suitable for installation on chassis with wheelbases from 120" to 136".

STANDARD EQUIPMENT:

Single rear door hinged at the right side over rear opening approximately 32" wide by 58" high.

Note the interior view showing a comfortable seat with rounded front to permit easy access by driver. The Urban Delivery is ideal for Department Store, Bakery, Dry Cleaning, and other House-to-House Delivery Services. Interior view shows unusual accessibility to both the driver's and the loading compartments.

The above view shows the Montpelier Urban Panel, Forward Control type, Model No. 11, 11 Ft. size, suitable for mounting on a ¾ ton, 1 ton or 1½ ton chassis, depending upon the load to be carried. This Unit is of unusual streamlined design and is adapted for House-to-House Delivery Service.

The above view shows the Montpelier Urban Delivery with its unusual streamlined design as adapted for Deluxe Department Store Delivery.

Prices and Specifications Subject to Change Without Notice

THE MONTPELIER MANUFACTURING COMPANY * MONTPELIER, OHIO

Specialized Motor Vehicles

A single page Montpelier Manufacturing sales sheet promoting the Model No. 9 and Model No. 11 Urban Delivery trucks. Although a specific chassis manfacturer isn't mantioned, this particular sheet was attached to a February 24, 1941 letter from the Chevrolet Motor Division to all direct Chevrolet dealers. Photos and text outlining the features of Montpelier's Model 323 and 423 Side-Aisle Delivery trucks was printed on the reverse side of this sheet.

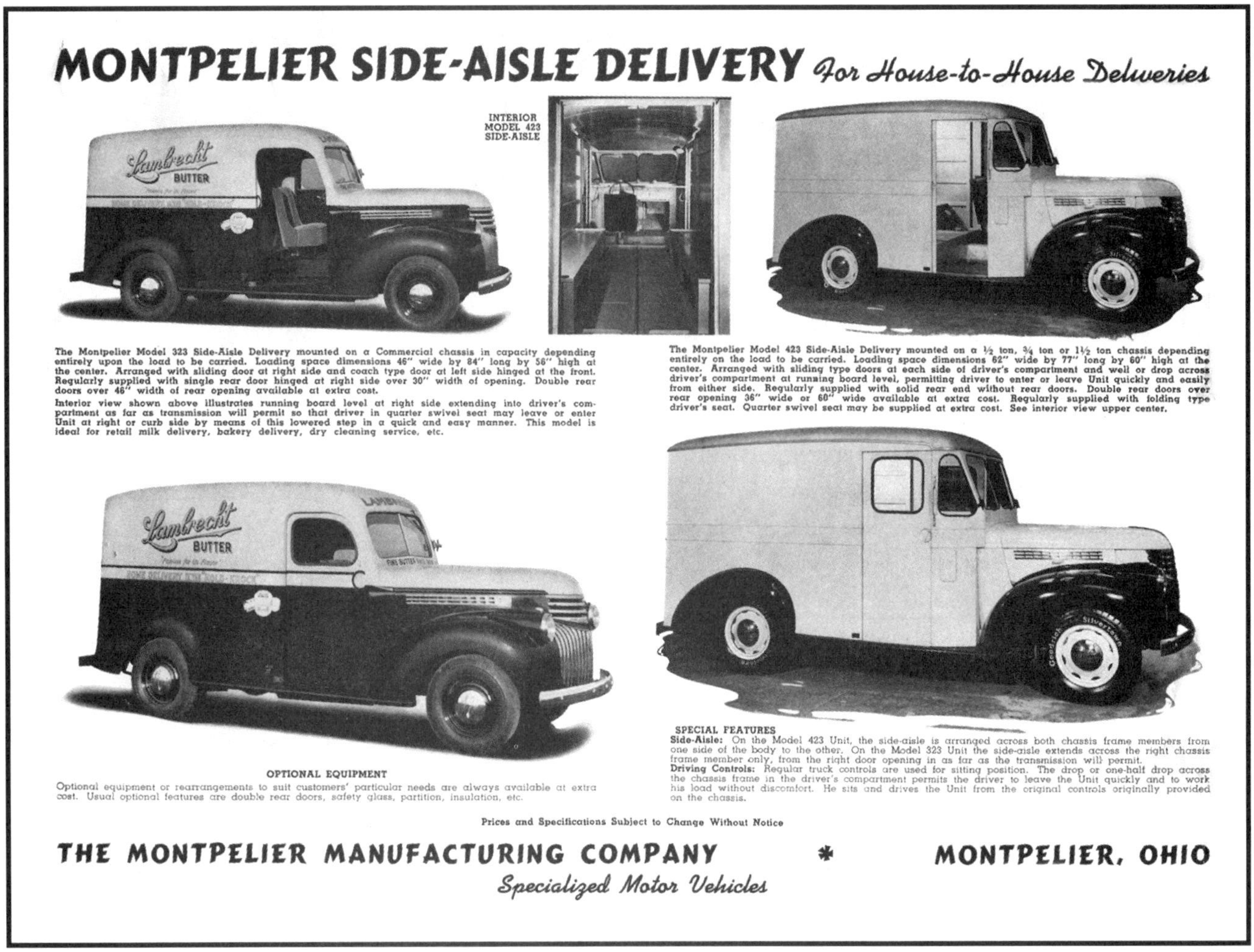

MONTPELIER SIDE-AISLE DELIVERY *For House-to-House Deliveries*

INTERIOR MODEL 423 SIDE-AISLE

The Montpelier Model 323 Side-Aisle Delivery mounted on a Commercial chassis in capacity depending entirely upon the load to be carried. Loading space dimensions 46" wide by 84" long by 56" high at the center. Arranged with sliding door at right side and coach type door at left side hinged at the front. Regularly supplied with single rear door hinged at right side over 30" width of opening. Double rear doors over 46" width of rear opening available at extra cost.

Interior view shown above illustrates running board level at right side extending into driver's compartment as far as transmission will permit so that driver in quarter swivel seat may leave or enter Unit at right or curb side by means of this lowered step in a quick and easy manner. This model is ideal for retail milk delivery, bakery delivery, dry cleaning service, etc.

The Montpelier Model 423 Side-Aisle Delivery mounted on a ½ ton, ¾ ton or 1½ ton chassis depending entirely on the load to be carried. Loading space dimensions 62" wide by 77" long by 60" high at the center. Arranged with sliding type doors at each side of driver's compartment and well or drop across driver's compartment at running board level, permitting driver to enter or leave Unit quickly and easily from either side. Regularly supplied with solid rear end without rear doors. Double rear doors over rear opening 36" wide or 60" wide available at extra cost. Regularly supplied with folding type driver's seat. Quarter swivel seat may be supplied at extra cost. See interior view upper center.

OPTIONAL EQUIPMENT

Optional equipment or rearrangements to suit customers' particular needs are always available at extra cost. Usual optional features are double rear doors, safety glass, partition, insulation, etc.

SPECIAL FEATURES

Side-Aisle: On the Model 423 Unit, the side-aisle is arranged across both chassis frame members from one side of the body to the other. On the Model 323 Unit the side-aisle extends across the right chassis frame member only, from the right door opening in as far as the transmission will permit.

Driving Controls: Regular truck controls are used for sitting position. The drop or one-half drop across the chassis frame in the driver's compartment permits the driver to leave the Unit quickly and to work his load without discomfort. He sits and drives the Unit from the original controls originally provided on the chassis.

Prices and Specifications Subject to Change Without Notice

THE MONTPELIER MANUFACTURING COMPANY * MONTPELIER, OHIO

Specialized Motor Vehicles

Montpelier's Model 323 and 423 Side-Aisle Delivery bodies installed on 1941 Chevrolet cowl and chassis units. Although Chevrolet chassis were used for this sales sheet, it's possible that this sheet was intended to represent Montpelier's capabilities for equipping the chassis of other truck manufacturers with similar Side-Aisle bodies.

International, Divco, White, GMC and Chevrolet *pushed* their somewhat generic delivery trucks out of the factory doors and into an extensive dealer network. With large advertising budgets, their trucks were heavily promoted in trade journal advertisements and elaborate sales brochures. The printed media extolled each and every design virtue, boasted of numerous fleets operating large numbers of their trucks, and hammered home the potential for greater economy, productivity and longer vehicle life.

In contrast, body builders like Montpelier lacked large advertising budgets and an extensive dealer network and, out of necessity, had to take the more challenging approach to survival by resorting to *pull* marketing. Dealing face to face and day to day with delivery truck fleet owners, offering them personal attention down to the finest detail and customizing their trucks to meet a fleet's exacting requirements was their formula for success. Their survival and success rested squarely on how persuasive their pull marketing efforts were and how well they delivered on their promises. The ability to sustain their manufacturing operations and their bottom line was often dependent upon whether or not Montpelier or one of its direct competitors were successful in winning a single contract bid for a few dozen trucks from a large metropolitan bakery, newspaper publisher, parcel delivery company, department store chain or dairy.

By the early 1940's side-aisle delivery trucks had evolved into a mature product. Changes from year to year were very small and beneath the skin, mostly confined to improvements to the cowl and chassis units that Montpelier and other body builders used as a starting point for this type of truck. Maintaining or gaining market share in the side-aisle segment of the delivery truck business was dependent upon providing a quality truck at a competitive price, backed up by responsive and highly personalized customer service. A body builder that failed to meet all of these criteria would likely be faced with the loss of business to a competitor who was a step ahead in one or more of these areas.

A different set of conditions applied to the forward control delivery truck business in the early 1940's. While not the first to introduce a forward control delivery truck to its customers, Montpelier's relatively early introduction of the forward control truck in the 1937 model year enabled it to achieve reasonable success as the 1930's drew to a close. By 1940, however, rapid engineering advancements to this comparatively new truck design were being made on an industry-wide basis and continued success for Montpelier's line of Urban forward control delivery trucks was in jeopardy. The Boyertown Auto Body Works of Boyertown, Pennsylvania was one of the earlier producers of forward control delivery trucks, starting in 1934. Boyertown's first forward control body was installed on a Diamond T chassis and was followed over the next twelve months by the production of roughly 100 similar trucks on all makes of chassis.

The Divco Model U, introduced in October 1937, was a truck that defied strict classification as either a side-aisle or a forward control vehicle so it successfully garnered sales in both segments of the industry. Not long after the Model U introduction, the Divco-Twin Company supplied variations of the Model U body to Chevrolet, GMC and Mack through the 1941 model year, enabling these companies to start or augment their sales of forward control delivery trucks.

Very large forward control delivery vans were in production at the Gerstenslager Company of Wooster, Ohio in the mid-1930's and by the early 1940's Gerstenslager bodies could be found on medium and heavy duty forward control chassis manufactured by GMC, Reo, Mack, Ford, Federal, Chevrolet, Studebaker, International, Dodge and others.

Reo, Willys and Hudson each entered the forward control delivery truck fray in 1939, 1940 and 1941 respectively, each in limited numbers and only for a brief stay in this intensely competitive market. Detroit's Federal Motor Truck Company also dabbled briefly in forward control delivery truck production during the same period.

International-Harvester's acquisition of the Metropolitan Body Company in 1938 was promptly followed by the introduction of the International Metro line of forward control delivery trucks in 1939. A formidable competitor, the Metro D-2-M and Metro D-15-M were the top selling forward control delivery trucks immediately after their inception and for many years thereafter.

White Motor Company introduced its White Horse line of forward control delivery trucks in 1939 and within a year or two was boasting of several fleets that were operating two hundred or more of them. The White Horse of this era was powered by an air cooled, horizontally opposed four-cylinder engine mounted just ahead of the transmission and rear axle. The entire drivetrain was a modular design that could be removed in its entirety for major repairs and replaced with a spare module to quickly get the truck back in service. The repaired module would then be a ready replacement for the next power module in a fleet of White Horse trucks that needed other than routine servicing. With one or two spare power modules on-hand and the ability to replace modules in a few hours or less, fleets could significantly reduce the number of spare trucks that were used as backups for trucks that were removed from day-to-day operation for repairs.

As was the case with International, General Motors quickly recognized the market potential for forward control delivery trucks and introduced their own Chevrolet and GMC forward control delivery trucks using bodies and chassis of their own manufacture for the 1940 model year.

Claims that forward control delivery trucks powered by internal combustion engines were first conceived and built either at Montpelier Manufacturing or the Boyertown Auto Body Works in the 1930's have to be set aside in light of at least one pioneering truck, the small but innovative Pak-Age-Car which made its first appearance as the Pac Car in 1925. As mentioned earlier, Stutz started marketing the Pak-Age-Car for the Package Car Corporation of Chicago in 1927, taking

REO's marketing department got way out in front of their engineering and manufacturing counterparts when they published a specifications sheet for this REO Dairy Car. Unfortunately, REO's 1939 efforts to enter the forward control delivery truck market never made it past the prototype stage. The main reason for the truck's debacle might have been the rather unorthodox drivetrain layout that had a four-cylinder engine mounted transversely at the front of the cargo compartment. The truck's radiator aperture is visible directly behind the driver's side jackknife door.

over production for the struggling Chicago company in late 1931 and reengineering the truck, bringing the body and drivetrain up to contemporary standards. After Stutz filed for bankruptcy in 1937, Pak-Age-Car tooling was acquired by the Auburn Automobile Company, a company that was no more viable in the late 1930's than Stutz. After filing for bankruptcy, Auburn's assets were reorganized as the Auburn Central Manufacturing Corporation and Pak-Age-Car delivery trucks trickled out of this meager company's Connersville, Indiana plant and were marketed through the Diamond T dealer network until 1941. Preceding the White Horse by fourteen years, the Pak-Age-Car also had a rear-mounted power module that could be removed and replaced with a spare module if major mechanical repairs were required. A Lycoming in-line four cylinder water-cooled engine served as the Pak-Age-Car's prime mover in its final years. White and others would adopt the removable power module concept in the three decades that followed the Pak-Age-Car's debut. The Pak-Age-Car's forward control configuration was a trend setting feature that will likely be with us for all of time.

The new forward control delivery trucks introduced by International, White, Chevrolet and GMC were clean sheet designs with little or no carryover from previous years of truck production. Steel was used for the understructure of their new delivery truck bodies and each company made sure that fleets and businesses in the market for delivery trucks were made aware of the fact in their sales efforts. It's certain that they pointed out that auto manufacturers had finally abandoned wood framing for car bodies by 1937 and that all-steel construction was indicative of the best and most modern method of construction for delivery trucks as well. Montpelier delivery bodies were still using quality wood framing with mortised, glued, screwed and ironed joints, a method that had served the company and their customer base well since the company's inception in 1927. By 1940 however, the cost of producing a wood framed body with an exterior steel skin and interiors of wood paneling was becoming a liability in the face of increasingly intense competition. Even if Montpelier's bodies held up as well as or even better than all-steel competitive bodies, the perception that Montpelier was offering a marginally obsolete design was getting difficult to avoid.

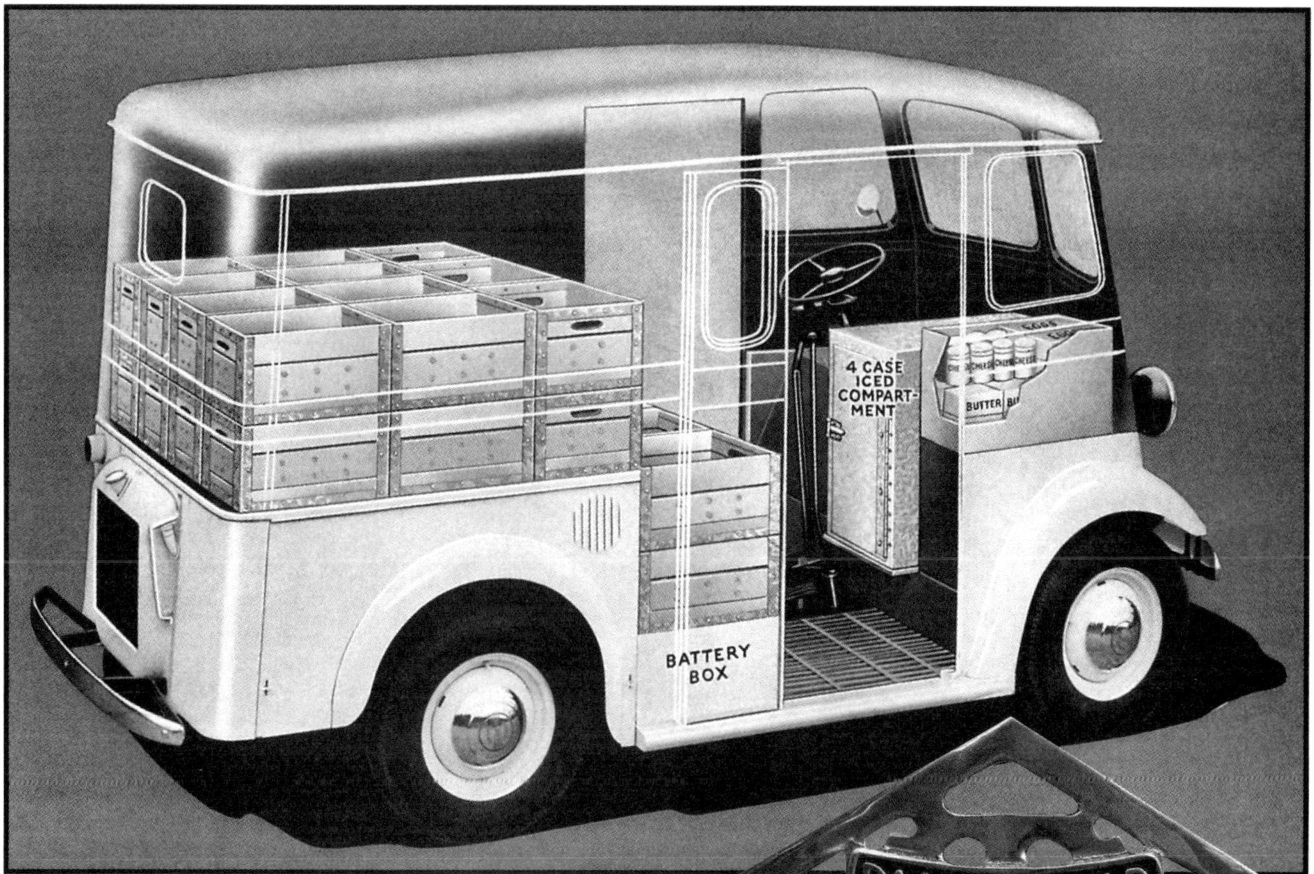

In its final design iteration, the Pak-Age-Car delivery truck was powered by a four-cylinder Lycoming engine and sold through the Diamond T dealer network. The rear-mounted engine was part of a complete power module that could be easily removed and replaced with a spare module if major mechanical repairs were required. Pak-Age-Car production began in 1925 and, with several major improvements, proceeded in limited numbers through the 1941 model year.

If the less obvious differences in construction methods weren't steering customers away from Montpelier's forward control delivery trucks, the more visible differences in the design of various delivery truck bodies from 1939 through 1942 may have been detrimental to Montpelier's sales. The new forward control trucks offered by International, Chevrolet and GMC featured smooth, expansive front end sections that did away with separately mounted front fenders. The smooth frontal appearance of these trucks gave them a modern look that was more compatible with the remainder of the truck body. Montpelier delivery bodies still employed separate front fenders, just as they did when first introduced in 1937. The separate fenders gave Montpelier's Urban delivery trucks a dated appearance by 1940 but for strictly practical customers who evaluated delivery trucks with stark objectivity, separate fenders were desirable for protecting the front corners of the truck body from damage and were easily and economically replaced if damage were to occur.

Prewar Montpelier Urban delivery bodies mounted on Dodge and Chevrolet chassis for the 1941 Model year can be found in this chapter. Chevrolet offered Montpelier Urban bodies on 3/4 ton, one ton and 1½ ton forward control truck chassis. Dodge offerings were limited to 3/4 and 1-ton chassis. Photos or promotional ads of competitive trucks from International, White, Divco, GMC, REO and Pak-Age-Car can be seen elsewhere on these pages.

The one-ton Special Delivery was one of three forward control delivery trucks fielded by GMC in 1940. GMC offered trucks similar to the one shown in this ad through the 1949 model year.

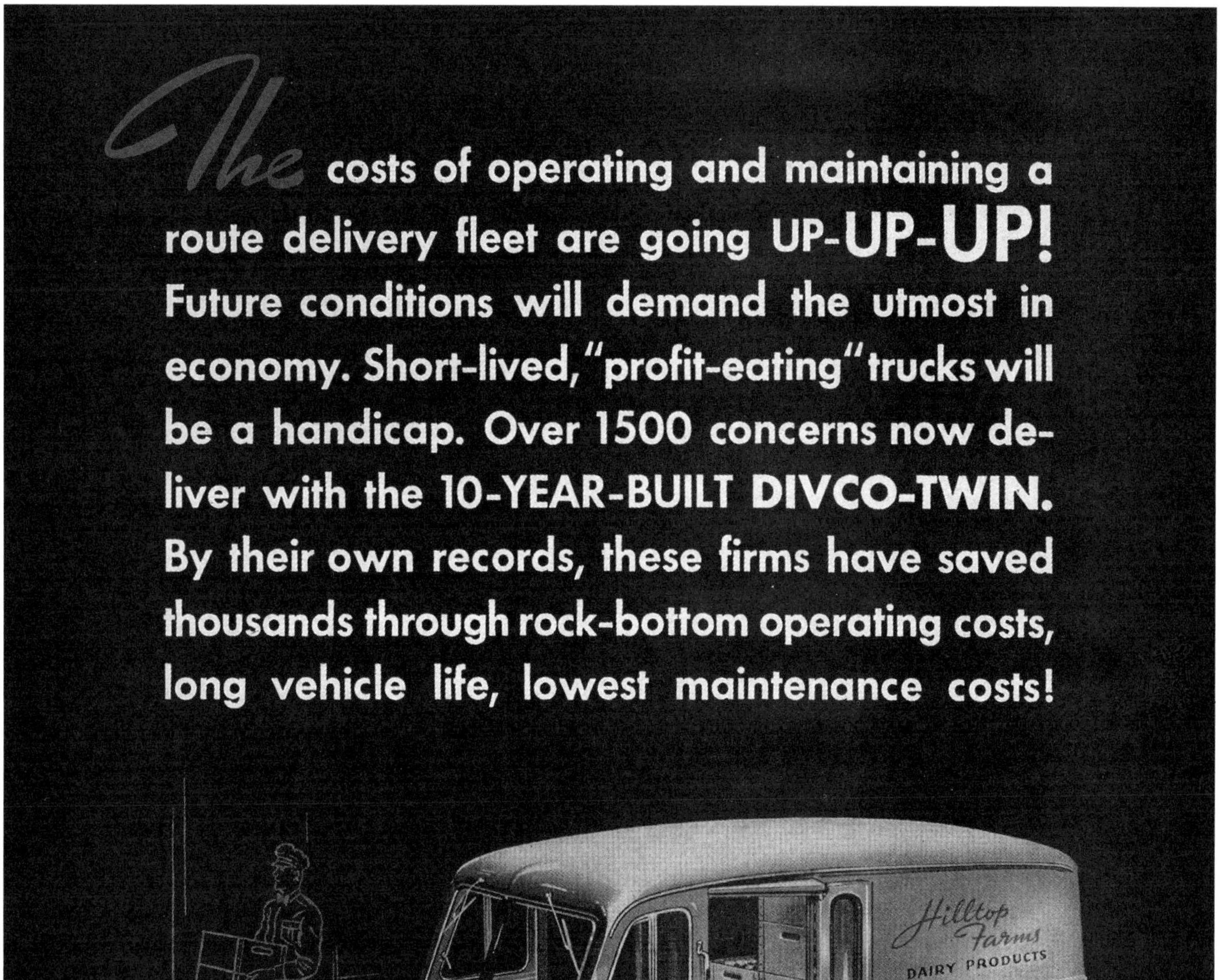

Starting in 1937, DIVCO delivery trucks like the one shown in this ad became a staple of the multi-stop delivery truck industry for a half-century. The 1941 model shown above looked much like the first DIVCO's as well as those that were produced until manufacturing operations ceased in 1986.

White Motor Company introduced the White Horse forward control delivery truck shown in this ad in 1939. An air-cooled horizontally opposed four-cylinder engine was part of a complete power module that could be quickly removed and replaced with a new module if major mechanical problems developed. Sales of the White Horse were brisk in its first few years of production but it was phased out shortly after World War II, an apparent victim of its own questionable design virtues.

This photo of Montpelier Manufacturing's office staff, shop superintendents and foremen (sans Harry Schwartz) was taken in the spring of 1941 or 1942, before a new office building was constructed in 1943. Standing from left to right are Bill Steinke, Bill Sprankell, Dwight Hostetler, T. H. Michaels, (person unknown) and Colin Cameron. Seated from left to right are Mary Wilkens, Ferman Haase, Bob Treer and Marcielle (Dick) Starr.

When faced with a dwindling order backlog, Harry Schwartz would embark on a week-long trip to Chicago where he maintained numerous customer contacts. Having sole authority within Montpelier Manufacturing to make on-the-spot deals, he would return with a fist full of new contracts, the lifeblood of the business for the next several months. Omar Bakery Company, headquartered in Omaha, Nebraska with bakeries and distribution centers in major midwestern cities, could often be counted on for a truck order during a Schwartz personal visit and a visit to executives at the Marshall Field department store often produced a fresh new contract as well.

Don Davis paneled truck interiors at Montpelier prior to World War II and often doubted Montpelier's ability to compete with an ever-increasing horde of competitors, in particular the high volume truck builders who were expanding into the forward control delivery truck market. His doubts were assuaged when he, as a shop worker, had the rare opportunity to talk to Omar Bakery representatives. Omar's experience proved that Montpelier bodied trucks survived up to three times longer than the bodies of competing trucks. Montpelier bodies were virtually unfazed by the rigors of daily operation on choppy city streets laced with potholes or

trolley tracks that were the bane of similar trucks offered by the giants of the U.S. automotive industry. Without doubt, Montpelier's contribution to the finished delivery truck was solely responsible for the truck's ruggedness and longevity since Montpelier and the high volume automotive giants both started with the very same chassis as a basis for their trucks. Lesser trucks were known to have high stress points that resulted in popped welds. The wood framing in a Montpelier body flexed just enough under severe service to absorb and distribute stresses over larger areas of the body structure instead of concentrating them in a few critical areas. It wasn't unusual for Omar to remove a Montpelier body from a broken down chassis and install it on a new or remanufactured chassis for a second tour of duty.

After suffering a stroke in April 1939, J. D. Hill passed away on May 7. Four days after his death, Harry Schwartz and Grant Stahl, representing a two-thirds majority of the remaining board members, elected Grace Hill, surviving wife of J. D. Hill, to serve as director and company president for the duration of J. D. Hill's unexpired terms.

To address the company's persistent financial problems, a May 20, 1939 board of directors meeting was called to discuss what was described a "new agreement" between Montpelier Manufacturing and "the owners of the mortgage deed on Montpelier Manufacturing Company property," i.e., Grant Stahl. Under the terms of the original land contract, Stahl had the option of foreclosing on the property if the company failed to make timely principal and interest payments. He graciously chose not to foreclose, apparently deciding that the local economy and the families of Montpelier Manufacturing employees would be better served by keeping the business in operation.

The new agreement called for the interest rate on the contract to be reduced from six percent to four percent, paid semi-annually each April and October. Four percent of the $23,100 principal would also be paid each year with the remaining principal to be paid off in its entirety on April 4, 1949. The agreement gave Stahl the option of declaring the full principal amount due and payable if the company failed to make timely payments of interest, taxes, assessments, liens and insurance premiums. It's apparent from these board meeting minutes that Montpelier Manufacturing had not made a single payment on the principal amount of the loan each April from 1929 through 1934 as specified in the original land contract.

The new agreement was short lived. Grant Stahl resigned as a director and company officer on July 11, 1939 and sold his shares in the company to Harry Schwartz (125 shares), Grace Hill (122 shares) and Margaret Hill Schwartz (three shares). While it's hard to discern from the rather sanitized minutes of the August 7, 1939 board of directors meeting, it's reasonable to conclude that Stahl resigned in exasperation, frustrated by the fact that his interests in the company were at variance with and would forever be subordinated to those of Harry Schwartz and Schwartz's mother-in-law, Grace Hill.

With the sale of Stahl's stock in the company, "Mr. Stahl agreed for himself, his heirs and his assigns to reduce the interest on certain notes due him from the company to 5% payable semi-annually; to accept repayment of principal on said notes at the rate of $200.00 per month until said notes are paid; and finally that any loans payable to either the National Bank of Montpelier or The Farmers & Merchants State & Savings Bank, Montpelier, which may be made from time to time shall have priority over his notes."

With Stahl's departure, Margaret Schwartz, wife of Director Harry Schwartz and daughter of Director Grace Hill, was elected to serve as a director and company Vice President and Treasurer. The composition of the board would remain the same for the next twenty years. A photograph of the board members, taken in 1944, can be found on the last page of this chapter.

Harry Schwartz initiated the next item of business at the August 7 board meeting. The meeting minutes noted that "it had been the custom of this firm each year to pay a salary to J. D. Hill for his services as President and Director.

"On May 7, 1939, Mr. Hill died, and as of that date no salary had been paid to him. Mr. Schwartz stated that the services of Mr. J. D. Hill were invaluable to this concern and that the loss of Mr. Hill would be felt for some time to come. It was further stated that the estate of J. D. Hill should be recompensed for the services of J. D. Hill for the period January 1, to May 7, 1939 and that said remuneration should also include in some small measure, a token of remembrance for the excellent service given this firm by Mr. Hill. Mr. Schwartz moved that a remuneration of $5,000.00 be voted in favor of J. D. Hill. This motion was seconded by Grace L. Hill and unanimously adopted."

A late August 1939 board meeting restructured the terms of the demand notes against Montpelier Manufacturing that Grace Hill inherited from her late husband. At a September 5 board meeting Grace Hill moved that "in consideration of the excellent service rendered by H. A. Schwartz during the current year, which has already resulted in a considerable increase in earnings over the previous year, a bonus of $5,000.00 be voted in favor of H. A. Schwartz for services rendered." The motion was seconded by Margaret Schwartz and unanimously approved. The restructuring of Grace Hill's demand notes was a sign that Montpelier Manufacturing was still struggling with indebtedness. One can't help but wonder if the payout to J. D. Hill's estate and the bonus accorded to Harry Schwartz would have been a source of bewilderment to Grant Stahl or Montpelier Manufacturing's other creditors if they had known about it.

A motion to issue 4,000 additional shares of stock was seconded and carried unanimously by all present stockholders, Grace Hill, Harry Schwartz and Margaret Schwartz, at an October 30, 1939 stockholders meeting. Board meeting minutes noted that "the capital stock of the Company is insufficient and it is necessary that same be increased to the amount hereinafter stated *(i.e., 4,000 additional shares)* for expansion, adjustment of financial structure and/or for other corporate purposes, as may be deemed necessary from time to time by the Board of Directors." For reasons unexplained, the three stockholders present at the meeting waived the right of current stockholders to purchase these new shares for a twelve-month period.

Moving directly from the stockholders meeting to the board of directors meeting, the three attendees elected themselves to the following offices: Grace Hill, President; Margaret Schwartz, Vice-President; Harry Schwartz, Secretary-Treasurer.

Only a week passed before Montpelier Manufacturing's board (with the exception, of course, of J. F. Robertson) reconvened. Shortly after the October 30 meeting, Harry Schwartz submitted an offer to immediately purchase 1,250 shares of the newly issued stock for $5.00 per share as well as the remaining 2,750 new shares at $10.00 per share on or before January 1, 1945. It's not certain why or how Schwartz was allowed to make a purchase offer in light of the limitations imposed on current stockholders at the October 30 directors meeting. Exerpts from the November 6 board minutes noted that a.) "it was desirable at this time to provide additional capital for the corporate needs of the company," b.) "H. A. Schwartz, acting as Secretary and General Manager of the company, has been largely responsible for the successful operation of the Company" and c.) "the aforesaid H. A. Schwartz has previously given and donated without compensation to the company sums totaling $6,000 to provide additional capital for the corporate needs of the company." The board directed the officers of the company to accept Schwartz's offer. So, Harry Schwartz essentially returned his August 1939 bonus of $5,000 to the company's coffers and, with an additional $1,250, acquired 1,250 new shares of Montpelier Manufacturing stock. Soon afterwards, he sold 450 of the newly acquired shares to his mother-in-law and president of the company, Grace Hill. As of January 2, 1940, Montpelier Manufacturing's stockholders were as follows:

Grace L. Hill	**822 shares**
J. F. Robertson	**250 shares**
Harry A. Schwartz	**1175 shares**
Margaret H. Schwartz	**3 shares**

In August 1940, the board voted that "in consideration of services rendered during the year that H. A. Schwartz be ***renum***erated *(sic)* in total amount of $15,000.00 and Grace L. Hill in total amount of $1,500.00."

J. F. Robertson, essentially an absentee director since resigning as a company officer in 1929, sold his shares in the company to Grace Hill and Harry Schwartz and submitted his resignation as a director on August 6, 1940. His resignation was accepted at the November 4, 1940 stockholders meeting and minutes of that meeting noted that shares in the company were now distributed as follows:

Grace L. Hill	**906 shares**
Harry A. Schwartz	**1341 shares**
Margaret H. Schwartz	**3 shares**

At the November 4, 1940 board meeting that followed the stockholders meeting, the board granted Harry Schwartz fifteen additional years on his offer to purchase the remaining 2,750 shares of stock that were issued by the company in 1939. Schwartz now had until January 1, 1960 to follow through on his offer to purchase the stock.

It's worth noting that, after Robertson sold his stock and resigned as a company director, Grace Hill, Harry Schwartz and Margaret Schwartz continued to serve exclusively as Montpelier Manufacturing's directors, company officers and stockholders for the next eighteen years.

At board meetings held on two successive days in May 1941, the board directed company officers to secure loans from The Farmers & Merchants State & Savings Bank and the National Bank of Montpelier totaling $17,500. Grace Hill signed an agreement "that all of the indebtedness of said Company to the above mentioned banks, shall be a prior claim over any indebtedness of said Company due to her."

At an August 4, 1941 board meeting, Harry Schwartz reported on the company's financial position and "called the attention of the Board to the unusual conditions under which the Company now is and for some time to come will have to operate. It was agreed upon by all present that the Company should continue its efforts to obtain work in connection with the National Defense Program." It was increasingly apparent that America's involvement with conflicts in the Far East and Europe was imminent and that the resources of automotive industry would be diverted toward government defense contracts and away from the production of truck chassis that Montpelier Manufacturing relied upon for its finished products. The board voted in favor of 1941 compensation levels for Grace Hill and Harry Schwartz that were identical to those granted at the August 1940 board meeting.

A 1944 photograph of the Montpelier Manufacturing Board of Directors and company officers, 1939 to 1959. Seated: Grace Stainthorpe Hill. Standing: Harry A. Schwartz and Margaret Hill Schwartz.

6

Montpelier Joins the War Effort

A welcome respite from the anxiety of constantly searching for fleet owners in need of new or replacement trucks came for Montpelier Manufacturing with the onset of World War II and numerous contracts that the company successfully won from the United States Ordnance Department, Army Air Forces and Navy. For Montpelier, these military contracts more than compensated for the virtual shutdown of commercial truck manufacturing during the war. Indeed, they brought the company its best years to date in terms of revenue and employment.

Harry Schwartz leads a groups of military dignitaries on a guided tour through the Montpelier Manufacturing plant. Civilian production of truck bodies had totally ceased and the entire plant was dedicated to high volume production of Army cargo bodies on three separate assembly lines.

At a January 5, 1942 board meeting, Harry Schwartz, "… called the attention of the Board to the effect upon the future of the Company of the general 'freezing' of the operations of the Automotive Industry by the Government." The board agreed with Harry's assessment that the company was capable of producing steel and/or wood fabrications required for the War effort and declared that every effort be made to obtain orders for any such items that could be produced with the Company's facilities.

When the war started, the company was at a loss as to how to support the war effort (and, in so doing, remain in business). Harry visited Wright Field near Dayton, Ohio and sought out military personnel that issued (or at least influenced the issuance of) military contracts. After explaining Montpelier Manufacturing's capabilities, Harry summed things up by stating that the company could make "anything that rolls on wheels." A colonel went over to a large table, peeled back a large cloth sheet that concealed scale models of numerous tactical vehicles and said, "Pick one, go design something for me and then come back and show me what you've got."

A three-quarter view of a WWII cargo truck body, one of thousands produced by Montpelier Manufacturing. The cargo body was a Montpelier Manufacturing design that, due to its outstanding design and construction, was adopted by all other cargo body builders at the direction of the War Department.

A finished and fully assembled cargo body, still on the cart that moved it from one assembly station to the next through the Montpelier Manufacturing plant. Loose items like top bows and tarps were never assembled at the plant as seen here. Instead they were packed inside the cargo body to facilitate transport by rail or truck to major 2½-ton and 5-ton Army cargo truck manufacturers.

By June 1942, Montpelier Manufacturing, under Schwartz's leadership, had "turned the trick," securing two lucrative government contracts, one for wood cargo truck bodies from the Quartermaster Corps and the other for catapults from the Army Air Corps. At a June 22 board of directors meeting, the board authorized company officers to borrow up to $350,000 to finance the manufacture of articles for the War Department. Montpelier Manufacturing was essentially requesting an advance on the government payments they would be receiving from the War Department through December 1, 1942. The advance would be paid off incrementally at an interest rate of 4½ percent as the War Department sent payments directly to the lender, the National Bank of Montpelier. A similar advance of up to $275,000 was obtained in February 1945 from three local banks under virtually identical terms.

Montpelier's body building expertise soon was amply demonstrated with the production of Army truck cargo bodies. Montpelier Manufacturing had the good fortune of being able to collaborate with the engineering division of the military service instead of the procurement division: This enabled it to develop, produce and sell an entirely new product that the procurement division may not have considered them capable of supplying. Military engineers were fully convinced that the Montpelier designed wood cargo body would meet or exceed the military's requirements and that Montpelier Manufacturing was more than capable of producing it.

Most former Montpelier employees estimate that over 50,000 cargo bodies were built by the company throughout World War II and into the mid-1950's. An undated but surviving document from the company's post-war files entitled "Partial List of World War II Contracts Completed, 1942-1945 Inclusive" accounted for the production of 38,321 bodies during the war. The document revealed that cargo bodies were manufactured in eight variations against twelve different Ordnance Department contracts. A scant three bodies were produced for trucks not intended for cargo or troop transport and twenty-five trailer-

PARTIAL LIST OF WORLD WAR II CONTRACTS COMPLETED 1942-1945 INCLUSIVE

Ordnance Department

Contract No.	Quantity	Product
W 33-019 ord-117	4800	Cargo Bodies, Wood, 80 x 144, Spec. ES-743C
W 303 ord-2518	1250	Cargo Bodies, Wood, 80 x 144, Spec. ES-743C
	500	Cargo Bodies, Wood, 80 x 108, Spec. ES-758B
W 33-019 ord-635	4000	Cargo Bodies, Wood, 88 x 204, Dwg. E-8617
W 33-019 ord-2944	1250	Splicing Kits for Wood Cargo Bodies, Dwg. D-7043962
W 33-019 ord-4129	8950	Cargo Bodies, Wood, 80 x 144, Spec. ES-743C
W 33-019 ord-4130	1561	Cargo Bodies, Wood, 80 x 180, Spec. ES-763B
W 303 ord-2431	750	Cargo Bodies, Wood, 80 x 144, Spec. ES-743C
W 33-019 ord-4511	3300	Cargo Bodies, Wood, 89 x 204, Spec. E-8617
W 33-019 ord-2665	8464	Cargo Bodies, Composite, 80 x 144, Spec. E-7064280
W 33-019 ord-3299	3422	Cargo Bodies, Wood, 80 x 108, Spec. AXS-957
W 33-019 ord-2911	1000	Cargo Bodies, Wood, 89 x 204
W 33-019 ord-3854	2980	Splicing Kits for Wood Cargo Bodies
W 33-019 ord-3853	100	Splicing Kits for Wood Cargo Bodies
W 33-019 ord-4431	321	Cargo Bodies, Wood, 88 x 168 Four Wheel Drive, Model 09426-Z
W 33-019 ord-4972	3	Bodies, Steel Truck, 88 x 240, 10-Ton 6x4 International
W 33-019 ord-4431	184	Blocks, Wood, Oxygen Cylinders, Dwg. A-339693
W 33-019 ord-5015		Wood Spacers - Machine Tools
W 303 ord-2519	25	Trailer to Bus Conversions

to-bus conversions were also manufactured. A copy of Montpelier's partial list of completed World War II contracts is found on these two pages.

A typical cargo body was produced to Specification No. ES-743C and was 144 inches long and 80 inches wide. With steel supplies extremely tight during the war, much of the cargo body was constructed of wood, including the frame rails. The use of steel was limited to crossmembers, the rails that formed the body's perimeter, stake pockets, bench supports, and miscellaneous gussets and braces. Wooden stakes were joined to arched wooden top bows by rectangular cross section steel tubing that was formed in a radius. Troop seats that were hinged to the truck body interior sidewalls could accommodate 16 combat equipped troops or, when folded down against the sidewalls, two and a half tons of cargo. The Montpelier factory had an excellent mill room that produced most if not all of the oak understructure for its Side-Aisle and Urban delivery truck bodies. However, wood was used so extensively in the manufacture of cargo bodies that the mill room was incapable of meeting production demands, so virtually all of the wood components were purchased from sources outside the company, freeing up former mill room workers to perform more vital production tasks.

U. S. Army Air Forces

Contract No.	Quantity	Product
W 20-017 ac-369	82	Catapults, Type A-1, Spec. 40620
W 535 ac-32761	96	Catapults, Type A-1, Spec. 40620
W 535 ac-29499	40	Catapults, Type A-1, Spec. 40620
W 535 ac-39910	120	Catapults, Type A-1, Spec. 40620
W 33-038 ac-7022	10	Catapults, Type A-2, Spec. 40720
W 33-038 ac-5757	16	Catapults, Type A-2, Spec. 40720
W 33-038 ac-8138	30	Catapults, Type A-2, Spec. 40720
W 33-038 ac-9048	20	Catapults, Type A-2, Spec. 40720
(33-038) 45-4191E	1	Catapult, Type A-3
W 33-038 ac-8643	40	Catapult, Type A-4, Spec. XR40937
(33-038) 45-3125-AF	1	Catapult, Type A-4, Experimental
(33-038) 46-2610	1	Catapult, Type A-5, Aerial Target
(33-038) 45-12411-E	1	Catapult, Type A-5, Aerial Target
(535) 43-6340-E	1	Catapult, Experimental
W535 ac-38527	100	Machine Gun Mounting Kits
W535 ac-34901	150	Machine Gun Mounting Kits.
(535) 42-16234	1	Hoist, Maching Set, Model 3642
(535) 42-6651	12	Tow Bar Assemblies, Airplane
(535) 43-13457-E	3	Scale Model Maintenance Ladders
(535) 43-671-E	5	Catapult Cars, Type OQ-2A
(33-038) 44-4800-E	6	Catapult Cars, 440-13750

United States Navy

Contract No.	Quantity	Product
N288s-26630(P)	5	Catapults, Type A-2, Spec. 40720
N288s-28386	14	Catapults, Type A-2, Spec. 40720

A recent photo of a Studebaker 2½-ton WWII cargo truck with a cargo body of the type produced by Montpelier Manufacturing.

The sheer volume of cargo bodies manufactured by Montpelier overshadowed all of the work the company had done in their fifteen-year history prior to the war. Struggles to keep the plant at full production ceased and employment grew to 200 workers on two ten-hour shifts. Ferman Haase, Montpelier's superintendent over steel fabrication and body assembly, would come home for dinner each evening at 5:00 and either return to the factory and continue working for another four hours or scour the countryside looking for potential employees to fill the company's constant need for additional help.

Frank Haase, Ferman's twelve-year-old son, would often make himself available to Carnice Kellogg to help assemble cargo bodies. (Carnice wasn't terribly enamored of his first name and very few people actually knew what it was. He insisted on using his last name as his sole moniker and that's what everyone called him.) Young Frank would keep a wrench affixed to the heads of bolts on the upper side of the cargo bed's slatted floorboards while Kellogg tightened the nuts from underneath. Kellogg paid Frank a dollar for each night's work, a tidy sum for a boy his age. Kellogg, however, was the chief beneficiary since he was paid on a piece rate basis and Frank's help enabled him to more than double his work output. For similar reasons, young Frank was also welcomed in the machine shop where Maurice Henry taught him how to run the Monarch lathe and a surface grinder.

Studebaker trucks "take to the air" in the service of the Red Army

THE realistic painting you see above was made from a photograph sent to Studebaker with the compliments of the General Automotive Division of the Red Army.

It shows how Studebaker trucks in the service of the Soviet Russian forces are swung, fully loaded, across unbridged ravines and streams.

The latest information obtainable indicates that a substantial portion of the powerful 2½-ton 6 x 6 trucks shipped to Russia from the United States have been Studebakers. These big Studebakers comprise approximately 40% of the hundreds of thousands of American trucks of all sizes made available to Marshal Stalin's fighting men under lend-lease.

In recognition of this important help, Red Army authorities presented Studebaker with an album of hitherto unpublished photographs of the trucks in action—"as a token of appreciation of the excellent quality of your military-type truck."

Supplementing its huge production of heavy-duty trucks for the armed forces of the U.S.S.R. and other United Nations, Studebaker to date has built over 60,000 of the Wright Cyclone engines that power the world-famous Boeing Flying Fortress.

And now an amazing new vehicle, the sure-footed, versatile Studebaker Weasel personnel and cargo carrier, is reaching America's far-flung fighting fronts in impressive quantities.

Awarded To All Studebaker Plants

Your War Bonds help America's fighters

Pictured is a convoy of Studebaker Weasels rolling inland from a landing beach. Remember, wherever they are—on land, at sea, or in the air—our valiant fighting forces need the help that your War Bonds give.

Montpelier Manufacturing's success in acquiring government contracts to support the war effort stretched the company's production capacity like nothing had before and brought unprecedented profits. Recognizing the need for supervision and officers to work harder and longer to meet delivery commitments, company directors authorized "that the Officers immediately arrange for the payment to supervision including officers and to office personnel, in such amount, and at such times as said Officers might deem it advisable, bonus in form of United States Bonds as extra compensation for the overtime and extra effort required during this trying period." Conspicuous by their exclusion from this authorization were the shop workers who were hard at work manufacturing the products in support of the war effort.

A resolution adopted by the stockholders at their annual meeting on October 26, 1942 authorized the Board of Directors to pay off as much mortgage indebtedness as earnings would justify in order to improve the company's financial position in preparation for post-war conditions. The directors, meeting immediately afterward, decided to pay off the first mortgage, eliminating the company's financial indebtedness to Grant Stahl.

As the books were closed for fiscal year 1943 on September 30, the War Department had determined that Montpelier Manufacturing had made excess profits of $31,000 on its defense contracts. Excess profits from defense contracts were a common occurrence among defense contractors during World War II. The priorities of waging successful campaigns in Europe and the Pacific were first and foremost in the eyes of not only the federal government but also American citizens and defense contractors. Montpelier Manufacturing's board of directors acknowledged the excess profits and the sum was entered on its 1943 tax returns and remitted to the Bureau of Internal Revenue by a December 15 deadline.

Montpelier's board of directors met on November 1,1943 and pointed out that, as stockholders, they had been denied dividends over the years when operating profits were used to make capital improvements to the plant. With surplus profits over and above liabilities and stated capital, the board approved a measure to grant each stockholder one-third share of unissued stock for each share of issued stock currently owned by the stockholders. To establish the value of these shares, Harry Schwartz, Company Treasurer, was directed to transfer $50,000 from the company's surplus profits to its stated capital and to issue new stock certificates for these shares. The release of 750 new shares, combined with 2,250 shares currently in possession of stockholders brought the total number of issued shares to 3,000.

Minutes of the October 23, 1944 annual stockholders meeting reflected that there indeed were 3,000 shares outstanding and that they were distributed among the three stockholders as follows:

Grace L. Hill	**906 shares**
Harry A. Schwartz	**2044 shares**
Margaret H. Schwartz	**50 shares**

In January 1944, Montpelier Manufacturing obtained a $395,000 line of credit "for the purpose of financing the manufacture by the Company of articles covered by war contracts, to repay existing indebtedness, to purchase U. S. Tax Saving Notes, and to free working capital upon cancellation of contracts." The line of credit was arranged with three local banks with the Cleveland Federal Reserve Bank serving as guarantor. Payment for the cargo bodies by the Army was sent directly to the loaning banks to reduce the principal amount of the loan rather than being sent to Montpelier Manufacturing.

In May 1944 the time once again came to reach reconciliation with the United States government on excess profits related to Montpelier Manufacturing's war contracts. Excess profits for the fiscal year ending September 30, 1943 were determined to be an amazing $560,000. Montpelier Manufacturing's board of directors authorized the company treasurer to pay the sum in two equal installments (less any tax credits or allowable deductions) to the Bureau of Internal Revenue on June 15 and September 15, 1944. Additional "renegotiations" in February 1945 and April 1946 prompted the board of directors to authorize the return of $640,000 and $207,788 in excess profits to the federal government for the fiscal years ending on September 30, 1944 and 1945.

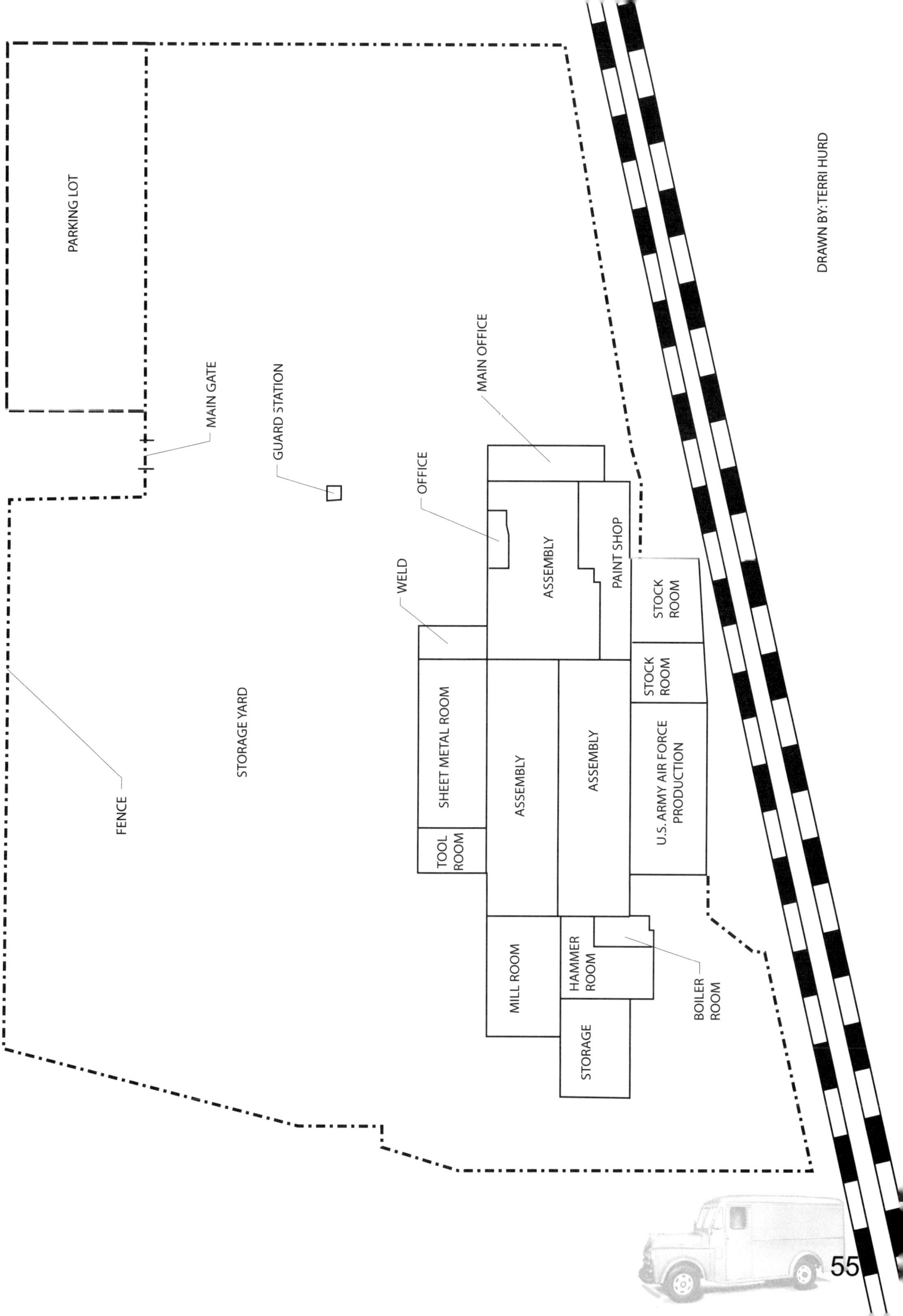
PARKING LOT
MAIN GATE
GUARD STATION
MAIN OFFICE
OFFICE
WELD
ASSEMBLY
PAINT SHOP
STOCK ROOM
STOCK ROOM
SHEET METAL ROOM
ASSEMBLY
ASSEMBLY
U.S. ARMY AIR FORCE PRODUCTION
TOOL ROOM
STORAGE YARD
FENCE
MILL ROOM
HAMMER ROOM
BOILER ROOM
STORAGE
DRAWN BY: TERRI HURD

This electric motor driven winch retracts the catapult's launching platform and projectile against resisting forces provided by six substantial elastic bands. Speed reductions obtained via the pulleys and a spur gear set increase the torque applied to the takeup spool in the top center of the photo. Upon releasing the trip lever, seen projecting through the safety shield, the elastic bands had sufficient energy to consistently propel a 150 pound object over 300 feet.

The Navy used these catapults, designed and produced by Montpelier Manufacturing, to launch radio controlled planes that, once airborne, were able to fly under their own power. If all went well, the plane's mission was short-lived, as their sole purpose was to provide targets for World War II Navy gunners.

Montpelier's labor force worked tirelessly and prospered as they manufactured cargo bodies. These jobs paid very well, as the piece rate basis was a lot less restrictive than it had been prior to the war. The large orders enabled workers to produce large quantities of parts and fully assembled bodies without having to dismantle manufacturing setups and convert fixturing to produce the wide variety of lower volume parts to fill random truck orders as they had in the past

Once completed, one cargo body was turned upside down and nested atop another. Loose pieces (top bows, stakes, tarps, etc.) for both bodies were packaged together and placed within the enclosure formed by the pair of bodies which were then strapped together. Montpelier Manufacturing hired Earl Wolff, the area Packard dealer, who, with the aid of his tow truck, hoisted the cargo bodies onto rail flatcars arrayed along the plant's south property line. The bodies were shipped to various military installations or truck plants where they were installed on trucks produced by every medium and heavy-duty truck manufacturer in the United States. Montpelier bodies could be found on any of these trucks but the majority were probably mated to some of the 562,740 GMC CCKW 2½ ton 6x6 trucks that performed a wide variety of missions during the war.

Montpelier Manufacturing produced other products for the Air Corps and Navy but the details had to be kept in strict confidence until the war's successful conclusion. By late 1945, the confidential nature of Montpelier Manufacturing's other war contracts no longer existed and residents learned that the company also had made catapults for the Navy and Air Corps. The catapults had a tubular frame, 30 to 40 feet long, supported by a single axle assembly. Two tracks started at the rear end of the catapult and terminated at an air cylinder that acted as a snubber, two-thirds of the way toward the opposite end. A winch retracted the projectile and its launching platform against resisting forces provided by six substantial elastic bands. Upon releasing the trip lever, the stored energy in the elastic bands would consistently propel a 150-pound block over 300 feet. The Navy used the catapults to launch radio controlled planes that, once airborne, were able to fly under their own power. If all went well, the plane's mission was short-lived, as their sole purpose was to provide targets for gunner practice. Harry Schwartz entertained the Montpelier Rotary Club in late December 1945 with films released by the Navy showing how the catapults performed these highly useful military functions. Catapult development testing taught Ferman Haase a few painful and frightening lessons that are recounted in his brief biography at the end of this book.

The roving government inspector responsible for insuring that Montpelier Manufacturing was fulfilling its quality and delivery commitments for World War II Navy catapults was none other than Maury Rose, the winning driver of the 1941 Indianapolis 500 race. After the war, Rose went on to become a three time Indianapolis 500 winner with successive victories in 1947 and 1948. During his visits to Montpelier, Rose, Harry Schwartz, Ferman Haase and Ferman's son, Frank, would enjoy a leisurely lunch at one of Montpelier's downtown eateries. It was quite a thrill for twelve year old Frank to rub elbows with a celebrity of Mr. Rose's stature and listen to stories of his racing exploits while pocketing a few of the dimes that Harry Schwartz gave him for musical entertainment via the restaurant's juke box.

Shortly after the war department authorized production, the initial prototype cargo body submitted to the war department by Montpelier Manufacturing was put on permanent display in the Pentagon.

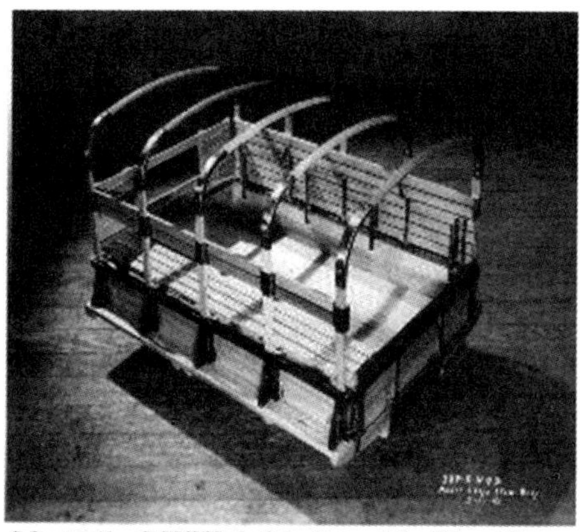

Montpelier CCKW. 6x6 scale model - Wood Cargo Body, primary design for the War Department, standardized February 1942. Cargo bodies built by a consortium of manufacturers. The Montpelier Manufacturing Company produced over 50,000 wood/composite steel cargo bodies during WWII. Army-Navy "E" Production Award - Two Stars

This is a photo of the scale model cargo body that Harry Schwartz presented to the engineering division of the War Department, seeking contracts for full scale volume production.

7
The Big Red "E"

It took Montpelier Manufacturing only seven months to distinguish itself as one of the Ordnance Department's top suppliers of cargo bodies. A September 3, 1942 *Leader-Enterprise* article pointed to reports from government inspectors that the outstanding materials and workmanship typical of the Montpelier bodies were far and above the inferior materials and poor construction methods of other manufacturers producing the same bodies as part of a military supply consortium.

Montpelier's consistent dedication to quality of design and manufacturing over the next two years led to widespread acclaim and recognition by the War Department and the various branches of the armed service. The culmination of this effort came with an announcement in the August 3, 1944 edition of the *Leader-Enterprise* that "the coveted Army-Navy 'E,' highest award given by the War Department for 'great accomplishment in the production of war equipment,' has been accorded the Montpelier Manufacturing Company."

A standing-room-only crowd assembles for the 3:30 PM ceremony honoring Montpelier Manufacturing for their steller contribution to the was effort.

Robert Patterson, Under Secretary of War, addressed the men and women of Montpelier Manufacturing in a letter that said, "This award symbolizes your country's appreciation of the achievement of every man and woman in the Montpelier Manufacturing Company. It consists of a flag to be flown above your plant and a lapel pin which each of you may wear as a sign of distinguished service to your country. I am confident that your outstanding record will bring victory nearer by inspiring others to high achievement."

The award recognized $8,000,000 worth of war contracts completed by Montpelier Manufacturing since the company dedicated itself exclusively to the manufacture of military cargo truck bodies in early 1942. August 22 was selected as the date for special ceremonies in Montpelier when company awards and a burgee type flag were to be presented. Lapel pins would be presented to each employee that was on the company's payroll before July 29. Since each employee played an integral part in the company's achievement and employees hailed from all parts of Williams County, Schwartz felt that it was only fitting that the entire county be asked to participate in the award ceremony.

At first overwhelmed by the rare distinction that was coming its way in a little over two weeks, it wasn't long at all before Montpelier Manufacturing and the citizens of Montpelier started firming up plans for the extraordinary day of recognition. The Montpelier Chamber of Commerce urged all business offices, stores and factories to close prior to the ceremony, thus giving all workers and citizens in the community the opportunity to attend. Harry A. Schwartz placed John C. Rundell in charge of preparing for the ceremony, sending him to the Cleveland Ordnance district office to finalize details. The ceremony would be conducted on a specially constructed platform on the north side of the Montpelier Manufacturing facility. Rundell arranged to have the platform appropriately decorated. Accommodations for two thousand people were provided. An assortment of Army and Navy equipment was put on display in the area surrounding the platform, including an M-4 tank, a "duck," a large armored car with a tank trailer and 12-foot and 17-foot trucks equipped with bodies produced by Montpelier Manufacturing.

Harry Schwartz praises the untiring efforts of all Montpelier Manufacturing employees in meeting delivery commitments and for maintaining exceptional quality standards in the production of WWII cargo truck bodies for the United States Army. Flags of all the Allied countries serve as a backdrop for the large and elaborately decorated platform that was built under the direction of event planner John Rundell.

Harry Schwartz and Dale Wisman accept the Army-Navy "E" Award Flag on behalf of the management and shop workers of Montpelier Manufacturing. All commerce within Montpelier was brought to a halt for the ceremony with the lone exception of the locomotive in the local rail swithyard, as evidenced by the telltale smoke plume behind the American flag in the back ground.

The Big Day, August 22, 1944

With the American flag raised by the local American Legion Post, more than 500 attendees opened the ceremony at 3:30 by singing "America," accompanied by the Paramount Band from Toledo. Rev. Forrest B. Brown, pastor of Montpelier's Methodist Episcopal Church, gave the invocation.

Lt. Floyd R. Mannon, second in command at the Toledo Naval Training Station, represented the Navy in presenting lapel emblems at the Army-Navy "E" presentation to employees of Montpelier Manufacturing Company. Representing the Army Ordnance Department was Major David Racoosin from the office of the chief of ordnance in Detroit. Harry Schwartz accepted the flag on behalf of the management of Montpelier Manufacturing. Accepting on behalf of the assembly department was Dale Wisman.

Master of Ceremonies Perry Faulkner delivered an introductory address from a speaker's platform adorned with American flags. Local dignitaries and special out-of-town guests occupied the platform with him. Faulkner was Montpelier's most civic minded and celebrated citizen, Ohio's Assistant Secretary of State and soon-to-be appointed to a federal post as head of the Veteran's Employment Service in Washington, D.C. "All of us," he stated, "are made to feel better by the knowledge that our citizenship is making a mighty contribution to the nation's gigantic effort for war."

Racoosin explained that Montpelier Manufacturing was the first plant of its type in the nation to receive an Army-Navy "E" Award. He then presented the "E" flag to General Manager Schwartz and Dale Wisman who, in turn, passed the flag along to the color guard which posted it at the right end of the platform, opposite the American flag.

Speaking for all Montpelier Manufacturing employees, Mr. Schwartz stated, "The Company takes but small credit for the honor given us. The honor comes to us because the men and women associated with the firm have worked earnestly and diligently to give the

War Department the material it needed. Many of the men and women have members of their families in the service and it's a matter of personal pride with them that the products we manufacture must measure up to only the highest standards of excellence. The Army-Navy production award that is being conferred upon our firm could not have been deserved or won except for the untiring loyalty and extra hours of hard work that the workers have given to their jobs. It is a real satisfaction to all of us to know that these efforts have been appreciated by our armed forces and that the products that we manufacture are so essential in helping to win the war. Likewise, this award is a challenge to each and every one of us to keep up our production with our very best efforts so that we may not fail to support our fighting men."

Montpelier Manufacturing employees identified by an August 17 *Leader-Enterprise* article as Schwartz's support staff included:

Materials: Lyle Rothenberger

Production: W. D. Sprankell and Ferman Haase

Purchasing: Robert Treer

Government Office Procedure: Marcielle Dick

Engineering: Kenneth Harman

Office Staff: Donna Ann Miller, Eileen Ensign, Bethel Brim

Bookkeeping: Mary Rummel

Shop: George Lamberson, T. H. Michaels, William Steinke, Robert Boszor, Dwight Hostetler, Wayne Bordner, Verlyn Heller, Harley Cramer, Jay Ransom, and Frank Smart

Extraordinary silence prevailed in the brief moments while Mrs. Hill received the first "E" emblem. Each and every person in attendance was completely absorbed as they grasped the magnitude of this awesome, unprecedented and yet fleeting moment in time. With sudden realization

Montpelier Manufacturing employees were seated in the first several rows directly in front of the podium.

Each active Montpelier Manufacturing employee that was hired on or before July 29, 1944 received an Army-Navy "E" award lapel pin at the August 22, 1944 ceremony that honored the company for its outstanding achievements in support of the war effort.

of the glory and honor of the award, tumultuous cheers and overwhelming applause shattered the silence as the crowd shared in the pride and national distinction that befell the men and women of Montpelier Manufacturing.

Grace L. Hill, Montpelier Manufacturing's president, and Albert Meiser of the plant's Air Corps department received the citation and the presentation of the "E" emblems from Lieutenant Mannon. Assisting him was Pfc. Robert Steffee of Ravenna, Ohio, a convalescent soldier who was wounded on D-Day on the beaches of Normandy with the first wave of American soldiers who began the invasion of France. His participation was a feature of the ceremony which had great and yet humbling emotional impact on those in attendance.

Chairman Faulkner concluded remarks made from the podium with a brief but heartfelt tribute to J. D. Hill, Mrs. Hill's late husband, past president of Montpelier Manufacturing and one of the company's founders. The ceremony closed with the singing of "The Star Spangled Banner." Some Montpelierites were fortunate enough to get rides in the tanks and other tactical vehicles and, although a fair amount of damage was done to Montpelier's streets by the tracked vehicles, the damage in no way detracted from the festive nature of the event and the overwhelming pride that saturated the town and permeated throughout all of northwest Ohio. Seven-year-old Jerry Schwartz was fortunate enough to get a ride in the M-4 tank. The tank operator, eager to impress his guests and enhance their once-in-a-lifetime experience, demonstrated the tank's "go anywhere" capabilities by taking down a tree that he considered an expendable item for such a special celebration. The tank's occupants were awestruck and, understandably, so was the tree's owner.

Although Montpelier Manufacturing produced undisclosed material for the Air Corps, the "E" award was conferred primarily for the innovative design, outstanding quality and timely delivery of the wood cargo body. Special guests and citizens from the surrounding region combined to make the ceremony an unprecedented honor for the company and for the town itself. If ever Montpelier would enjoy a glorious day of fulfillment, pride and jubilation, this would be just that very day. At the presentation's end, it was apparent that much credit was owed to John Rundell for the countless hours spent in supervising the construction of the platform, decorations, publicity and overall coordination of each and every detail of the event.

A mere six months after the Army-Navy "E" Award was presented to Montpelier Manufacturing, the award was renewed by Undersecretary of War Robert P. Patterson with the presentation of a white star that was to be placed upon the E flag presented to the firm in August 1944. The public relations department at Wright Field took note of the award renewal and commented that it was quite unusual for a company to be recognized so quickly with a white star. Most companies receiving similar recognition waited from a year to eighteen months before receiving the highly esteemed white star award.

A special ceremony the likes of which occurred during the initial presentation of the "E" Award was not something the War Department engaged in, leaving celebration and recognition of the achievement up to the recipients. Harry Schwartz was away on business at the time the award was received and it is not known how the company celebrated this extraordinarily high achievement. Incredibly, another White Star was awarded to the company by Patterson in September 1945.

A selection of tanks and military vehicles paraded through the streets at the celebration. The resulting damage to streets and trees was deemed an acceptable for this extraordinary occasion.

The M-4 battle tank appears ready to take off with its first load of local citizens, soon to be joined by the "duck" at the right and the armored personnal carrier in the center of this picture.

At the war's outset, it appeared that Montpelier Manufacturing would not have the good fortune of finding a way to contribute to the war effort. However, Harry Schwartz correctly perceived that steel shortages would likely become a critical factor, forcing delays in delivery of critical materiel and, consequently, delays in prosecuting the war. Schwartz embarked on the design and construction of a wood cargo body, presented the design to the War Department and received enthusiastic approval and authorization to start production. The plant that only a short time earlier couldn't find a way to contribute to the war effort went on to become a principal producer of critical Army Ordnance supplies and, furthermore, received a distinction that less than three percent of all war-producing firms in the United Stated had earned. Of those companies receiving the "E" award, only a very few operated in communities as small as Montpelier. The cargo body produced by Montpelier Manufacturing was standardized by the Army, making it necessary for other contractors to supply cargo bodies to the specifications developed by Montpelier Manufacturing

Preparing for the Post-War Era

An engineering drawing of the plant layout in 1951 reveals that over 9700 square feet of floor space used during the war to assemble cargo bodies remained vacant within the plant in the hope that future defense contracts would be granted. Based on the merits of their Army-Navy "E" Awards, Montpelier Manufacturing was optimistic that more military work would be forthcoming in the years that followed World War II. The company was pleasantly surprised (and caught a little off their guard) in 1953 when a final large order for several thousand cargo bodies was received, forcing workers to desperately scavenge remote areas of the factory for old fixtures that were used to make various cargo body components. Production tooling was located and set up once again to produce cargo bodies in a dedicated area of the plant for the next two or three years. By the time this final order was received, the steel shortages of World War II and the Korean War had subsided and a greater amount of steel was used in cargo body production. Top bows, sidewalls and much of the cargo body framework were converted from seasoned oak to steel.

By the September 10, 1945 Montpelier Manufacturing board of directors meeting, government restrictions on the compensation of high ranking officers of companies holding war contracts had been lessened. Following the approval of an application to the Salary Stabilization Unit of the U. S. Treasury Department, Montpelier Manufacturing's board of directors rewarded Harry Schwartz for his leviathan efforts in designing an innovative military cargo truck body and catapult for the benefit of the United States Armed Forces (not to mention the economic survival of the company itself). The reward came in the form of $12,500 salary in arrears for each of the three fiscal years ending in September 1943, 1944 and 1945. Going far beyond the mere design of these defense products, Schwartz was also responsible for securing the government contracts to produce them and for leading the company's effort to convert its manufacturing resources from the production of civilian delivery truck bodies to cargo truck bodies and catapults. Winning the coveted Army-Navy "E" Award and the two follow-up white star awards provided ample verification of the incredible value of his strategic initiative in the eyes of the board members as well as the employees of Montpelier Manufacturing and their families. The supplemental salary for fiscal years 1943 through 1945 brought Schwartz's annual monetary compensation to $30,000. Effective October 1, 1945, Schwartz's salary returned to its previous level of $17,500.

Earned income for other Montpelier citizens during the war might help to put Schwartz's annual salary into better perspective. Annual salaries for various employees in Montpelier's local school system were as follows:

Grade School Teacher:	**$1,190**
High School Teacher:	**$1,375**
Grade School Principal:	**$1,825**
High School Principal:	**$2,250**
School Superintendent:	**$2,950**

8

A New and Expanded Line of Urbans

The Low-Step Urbans introduced shortly after World War II turned out to be a stopgap design for Montpelier. With the introduction of the new line of Urban delivery trucks in 1947, a change was instituted in the way these truck bodies were built. For Model 9 and 11 Low-Step Urban truck bodies as well as their predecessors, sheet metal workers were their own pattern makers. It was common practice for each sheet metal worker to make his own pattern for a certain body panel. Many of the body parts and panels simply were not made to a "hard design" blueprint. The more complicated and contemporary Urban models that appeared in 1947 made it necessary for the pattern shop to take pattern making skills to a higher level. It provided the sheet metal workers with more exact patterns that could then be copied and formed into body panels that weren't so highly variable from truck to truck.

A 1946 Montpelier Manufacturing brochure promoting features of the redesigned Low-Step Urban Delivery Bodies. With a fairly short life cycle, production of this forward control delivery truck design started immediately after World War II and was phased out in mid-1947. The back page of this brochure pointed out the features of Montpelier Side-Aisle Delivery bodies of the same era.

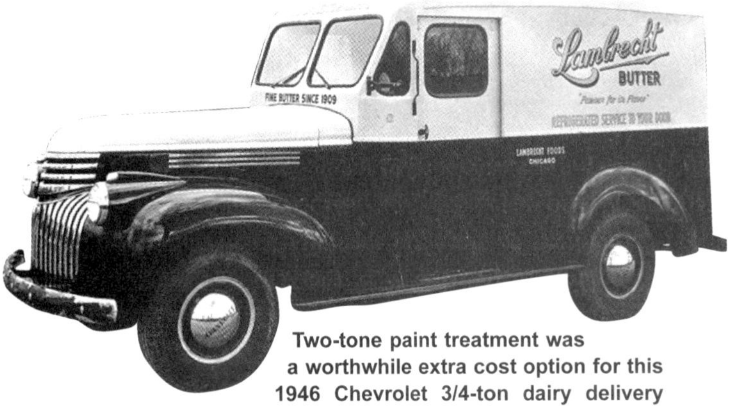

Two-tone paint treatment was a worthwhile extra cost option for this 1946 Chevrolet 3/4-ton dairy delivery truck equipped with a Montpelier Model 423 Side-Aisle body. Montpelier's extensive chassis modifications (wheelbase alterations, installation of the Side-Aisle drop box and stand-drive controls) aren't readily apparent in this photo.

A modified 1946 Chevrolet 1½-ton cowl and chassis equipped with a Montpelier Model 423 Side-Aisle delivery body just prior to being transported to a Montpelier customer for years of reliable daily delivery service.

A 1946 Chevrolet 1½-ton chassis provides support and motive power for this good looking Montpelier Model 11 Urban delivery body equipped with optional rub rails and rear wheel skirts. The attractive appearance of this truck undoubtedly gave the Banner Linen Service an image boost in daily delivery service.

Model designations for the new Urban bodies introduced for the 1947 model year were once again based on the length of the cargo compartment in feet, continuing the simple but logical precedent that was established when Urban Models 9 and 11 were released in the 1930's. Replacing the obsolete Models 9 and 11 were new Urban Models 8, 10, 12 and 15.

Beginning in 1949, Montpelier published specification sheets for the new Urban delivery bodies that were tailored to each of their chassis suppliers. The bodies were very similar from one brand of chassis to the next, the only external differences being hubcaps, minor variations in external bright trim and small variations in wheelbase length. With the debut of the new Urban models in 1947, all use of wood for the body understructure was abandoned. Plywood was retained for paneling Urban interiors (as indicated on the 1949 Chevrolet Urban spec sheet) and wood decking was used for the Urban's floors. The materials used for interior paneling weren't mentioned in the 1951 specification sheets. Only the Chevrolet specification sheet for 1951 makes mention of kiln dried hardwood flooring. It isn't clear whether Montpelier purposely decided not to call attention to the continued use of wood or if the company was in the process of changing to all steel construction as its competitors had.

Standard equipment on a 1947 Urban body included easily operated sliding type doors on either side of the driver's compartment, a single 32 inch wide windowless rear door, plywood paneling on interior side walls, a driver's compartment dome light, an outside rear view mirror, a driver's bucket seat on a swivel and forward tilt pedestal, safety glass in all windows, outside door handles with keyed locks, dual fresh air ventilators for the driver's compartment, square type rear wheel housings, a hinged, insulated and easily removable engine cover, a channel rear bumper with safety tread step plate, and a primed exterior finish. The interior walls were parallel in order to provide maximum useable cargo space and facilitate installation of racks or shelves.

Extra cost options included a fully insulated cargo compartment, folding side doors, four double rear doors options in varying widths, glass in rear doors, an auxiliary folding passenger seat, a partition with sliding door separating the driver and cargo compartments, a wrap around rear bumper and rub rails at the skirt line for improved side panel protection, an extra dome light, heater/defroster unit, running lights, directional signals and single or two-tone paint in the customer's choice of color.

A new process was instituted for forming the two panels that comprised the front clip on an Urban, from one side door to the other and from the cowl down to the bumper panel. Large flat sheet metal blanks produced in George Cornell's sheet metal department were transferred to Lee Henry in the hammer room. The blanks were spread over a die cavity and clamped in place with air-operated clamps. A hydraulically activated punch rose from beneath the sheet steel blank, stretching the sheet over its contours. After some edge trimming, the formed panel was ready to be welded to its symmetrical counterpart to form the complete lower front end section of an Urban delivery body.

A 1947 Model 8 Urban body mounted on a Chevrolet forward control chassis received a custom exterior paint treatment. It was photographed for promotional purposes and then used as a mobile demonstrator for Montpelier's complete line of contemporary forward control Urban delivery trucks. Two tones of green exterior paint contrasted superbly with white for a fresh new look unlike few Montpelier trucks before it. Chevrolet was sufficiently pleased with the truck's appearance to include a small image of it in their 1947 full-line truck brochure. Recalling the favorable reception that the display truck received, Montpelier Manufacturing used the same paint scheme ten years later when introducing a newer and more advanced forward control delivery truck, the Mono-Lite Model 75. At the time that the Mono-Lite Model 75 was introduced, International released a very similar but smaller forward control delivery,

A Montpelier Model 10 Urban Delivery body, when mated to a GMC P-152-22¾-ton chassis, resulted in a truck with 350 cubic feet of cargo volume on a 125¼ inch wheelbase. Overall length of the truck was 211 inches.

Lamson Brothers was a prominent department store in Toledo, Ohio. Like many department stores that delivered their customers' purchases, Lamson Brothers management stressed the importance of a delivery truck was clean and made a good visual impression on its delivery route as evidenced by these circa 1951 Montpelier Urban Model 10 Delivery trucks.

Montpelier installed their Model 12 Urban Delivery body on 1949 Ford F-5 1½-ton forward control chassis of 134 inch wheelbase to provide a truck with 420 cubic feet of cargo capacity.

Montpelier Side-Aisle and Urban bodies were supported and moved by carts as they progress though various body assembly stations. Soldered joints at the windshield divider and each "A" pillar are visible on the Urban body at the right. Framing is visible in an Urban body that awaits assembly of body side panels at the far end of the assembly line.

This 1948 Urban Delivery body mounted on a Chevrolet chassis received special three-tone paint treatment to call attention to Montpelier's new line of Urban deliveries. The same paint scheme was used on one of the first Montpelier Mono-Lite 75 trucks ten years later. International appropriated a very similar paint scheme in 1958 for its new Metro-Mite 80 compact forward control delivery truck in an apparent effort to divert attention away from Montpelier's new Mono-Lite truck.

the Metro-Mite 80. Rather brazenly, International appropriated Montpelier's exterior paint scheme in promoting the Metro-Mite 80 in what appeared to be a shameless attempt to divert the attention of prospective customers away from the Mono-Lite and toward the Metro-Mite 80.

Replacing the pre-war Model 323 and 423 Side-Aisle deliveries was a four-model lineup consisting of Models 405, 410, 415 and 420. Intended primarily for the delivery of dairy products, the Side-Aisle specifications sheet provided information on how many round or square milk bottle cases could be accommodated in the cargo compartments of the largest three models. The use of wood was totally shunned for these bodies; floors were steel, as were the interior sidewalls. The underbody and all interior and exterior steel panels were rust-proofed before assembly.

Standard features across the entire line of Side-Aisle deliveries included double rear doors mounted on piano hinges, a driver's seat that folded as well as tilted forward, low-step entry from either side of the driver's compartment, parallel interior sidewalls for maximum cargo space and easy installation of shelves or racks, a driver's compartment dome light, a shelf in the driver's compartment for route books or other day-to-day items, windshield and side door safety glass, outside door handles with keyed locks, dual fresh air ventilators for the driver's compartment, square type wheelhouses, a channel rear bumper with safety tread step plate and a primed exterior finish.

A one-of-a-kind show truck on a 1948 Chevrolet cowl and chassis unit, this Side-Aisle Delivery differs substantially from a production model with its aluminum cladding, jackknife side door with triangular windows arranged symmetrically around a rectangular window and steeply raked windshield. Note the gas filler cap atop the right front fender. Montpelier employees Leo Mansfield and Art Donaldson are standing by the overhead door in the background. The dark colored entry door is the door through which shop employees entered when reporting to work. The office entrance is partially concealed by the truck's windshield header.

Optional features available at extra cost included reinforced rear springs, dual windshield wipers, a bucket seat on a swivel pedestal, a fully insulated cargo compartment, full width double rear doors, rear door window glass, protective rub rails on interior walls and single or two tone exterior paint in the customer choice of color.

Montpelier produced a very interesting Side-Aisle "show truck" on a Chevrolet flat-faced cowl and chassis for the 1948 model year. Sharing virtually nothing in common with the more pedestrian looking standard production side-aisle deliveries, the one-of-a-kind custom truck sported front fenders that faired smoothly into the body side panels, extensive use of bright extruded aluminum cladding on the body and front fenders and, for lack of a better term, a "sporty" backswept windshield. The front corner window and the rearmost window on each folding side door were identical in shape and arranged symmetrically on either side of the remaining window in the folding side door. To avoid interrupting the smooth horizontal lines of the aluminum cladding along the truck's skirt, a special fuel tank was installed behind the right front fender, its filler cap visible just ahead of the Chevrolet Thriftmaster nameplate on the hood.

Envisioning a Side-Aisle delivery truck clad in the equivalent of a formal tuxedo enables one to appreciate the eye appeal this truck had, especially when it was parked beside one of its more plebeian brethren. This special side-aisle truck was built for display at one or more trade shows but records of its life in the limelight no longer exist. When the truck's tour of duty was completed, it was apparently sold to the Omar Bakery Company.

A Meadow Gold Side-Aisle milk truck from the 1952 model year. Including the customer's name in this promotional photo amounted to a testimonial from a large producer of dairy products. Montpelier hoped that endorsements from prominent customers like Meadow Gold would attract new business.

A circa 1950 Willys two-wheel-drive cowl and chassis supporting a Montpelier Side-Aisle Delivery body. Montpelier willingly utilized any chassis that a customer would specify but requests for Willys underpinnings were infrequent. Willys produced a rugged panel truck with at least some specifications that were comparable to a Montpelier Side-Aiale Delivery truck that probably sold at a much lower price than the unit illustrated here.

Of little or no functional value, cladding similar to that used on this show truck was never used on production Montpelier truck bodies. It was, however, adopted years later on a much smaller scale for decorative purposes on forward control delivery truck bodies produced by the Boyertown Auto Body Works, a Montpelier competitor located in Boyertown, Pennsylvania.

In light of increasing inventories and capital improvements made to its manufacturing plant, Montpelier Manufacturing's board of directors determined at a May 24, 1948 meeting that it was necessary to transfer $150,000 of its surplus capital to the stated capital represented by its outstanding shares of stock. In doing as it had done on at least two previous occasions, the board increased the value of its shares to be more closely aligned with the book value of the company.

Minutes of the October 18, 1948 stockholders meeting were the first of their kind to record how the 5,000 outstanding shares of Montpelier Manufacturing stock were distributed among its stockholders. The stockholders and their holdings in the company broke down as follows:

GRACE L. HILL	906 SHARES
MARGARET H. SCHWARTZ	50 SHARES
HARRY A. SCHWARTZ	4044 SHARES

Stock transactions had apparently taken place after the board declared the dividend payable at the rate of 2/3 share in previously unissued stock for each outstanding share of stock at the January 20, 1947 directors meeting. Both Grace Hill and Margaret Schwartz had either sold or transferred the dividend they received in new shares to Harry Schwartz.

An indefinite layoff affecting roughly one-half of Montpelier Manufacturing's work force was announced in late January 1949. Montpelier's order backlog generally dwindled during the winter months but this layoff was more extensive than any in recent years. The announcement, initially posted on company bulletin boards, indicated that the layoff was due to a shortage of available truck chassis and a dwindling backlog of truck orders. "Further," the statement read, "due to model change-over and a highly competitive situation, we do not see how production can be resumed at this time at the current production schedule." The announcement indicated that the layoff would remain in effect until the volume of available chassis and confirmed orders warranted an increase in production. "We have no knowledge, of course, as to how long this will take," the statement concluded, "but if we all work together to meet this situation, we feel that we can again face our competition on a fairly equal basis."

Frequent references to "this difficult period" or "the present difficult position" were rarely accompanied by supporting information in the minutes of stockholders or board of directors meetings, a notable exception being the annual stockholders meeting of October 31, 1949. Steel and coal strikes were hampering the company's operations and "It was noted that the company's order back-log has all but disappeared, and it was the general feeling that the coming year would be one where business would be very difficult and all possible economies were to be practiced". Undaunted at least to some extent, the stockholders adopted a resolution authorizing the board of directors to "pay dividends to stockholders to such an extent as may be justified in the opinion of the Board of Directors by earnings and prevailing economic conditions". The board of directors complied, indicating that the usual dividend of $5.00 per outstanding common share would be paid to stockholders of record on January 15, 1950, if business conditions should warrant.

Montpelier Manufacturing received a special order in 1949 that, unfortunately, was destined never to be repeated. The Truck Division of the Eisenhauer Manufacturing Company ordered five COE truck cabs from Montpelier for a limited-production run of prototype Eisenhauer X2 heavy-duty long haul trucks. Eisenhauer's Truck Division was located in Louisville, Kentucky and was incorporated as a separate company from its parent company, the Eisenhauer Manufacturing Company of Van Wert, Ohio. The Eisenhauer X2 was the Truck Division's second generation, twin-engine straight truck with a cargo capacity of twenty tons, achieved within an overall length of just thirty-five feet. Sporting many unique features in addition to its twin in-line six cylinder 302 cubic inch displacement GMC engines, the trucks also were equipped with twin steerable front axles and three rear axles, the front and rear most of which were drive axles with a dead axle in between. Rear axle suspension was designed in such a way that the two drive axles steered as well, pivoting in equal but opposite directions in a plane parallel to the ground as the truck progressed through a turn.

Eisenhauer twin engine trucks incorporated many thoughtful advancements that were geared toward benefiting truck owners, operators and even the nation's overtaxed highways and bridges in the years following World War II. However, in an environment with many parallels to the post-war Tucker automobile, its developers lacked the financial resources to tool up for high volume production and the steep challenge of competing head-to-head with well-established heavy-duty U.S. truck manufacturers. As a result, the trucks never progressed beyond the field test phase that was finally terminated in 1957. Montpelier Manufacturing was given the unique opportunity to build the COE cabs for these trucks, cabs that were every bit as impressive as the trucks' other design innovations

The sole topic of a January 20, 1950 board of directors meeting was the company's policy of awarding discretionary bonuses to monthly (i.e., employees paid a monthly salary rather than an hourly wage), hourly and production employees. The monthly employee bonus plan began in 1947 and was, "based on general business conditions and its earnings, paid to all of its monthly employees at three intervals during the year in such amount as in the opinion of management represented a measure of each monthly employee's contribution and service to the successful operation of the business." Company officers set aside

In 1950, Montpelier Manufacturing produced the cabs for five prototype Eisenhauer X2 trucks. These trucks featured twin GMC 302 CID in-line six-cylinder engines and twin steerable front axles. Two of the three rear axles were also steerable. This truck is equipped with a 6,176 gallon Fruehauf tank. The five prototypes were extensively field tested into the later 1950's but never progressed beyond the prototype stage.

ten percent of projected net earnings in order to provide monthly employees with this discretionary benefit, best described in today's terminology as profit sharing.

The board determined at this meeting, "that each hourly and production employee, starting on June 30, 1950, and each year thereafter, until withdrawn or modified by the Company, shall receive a service bonus in the amount of five dollars for each continuous year of continuous service in the employee of the Company, but with a top limit of $75.00 to any one such employee." Qualifying for this bonus might have been frustratingly difficult for hourly and production employees as they were subject to layoffs that occurred rather routinely during model changeover periods in the fall and in January and February when business was typically at a low ebb. Harry Schwartz was well known for implementing layoffs at the first hint of a business slowdown. Since the shop was not unionized, a layoff of shop employees occurred whenever and wherever it was feasible. The size of the labor force was scaled back during business slowdowns to match the volume of incoming orders with virtually no regard for employee seniority. Consequently, Montpelier Manufacturing employees never burned the bridges they left behind at previous employers. Many of them returned to work at an old job at one or more regular intervals during the year for anywhere from a few weeks to as much as six months.

"In light of prevailing economic conditions," the discretionary bonus plan for monthly employees was once again a topic at a November 24, 1950 board of directors meeting. Company officers found it difficult to determine the size of the overall bonus payment for monthly employees at three intervals during the year because of their inability to accurately estimate the company's net income after taxes. A review of the past four years of monthly employee bonus payments revealed that they had received over $63,000, or 6.4 percent of the company's net profits in the form of discretionary bonuses. The plan had worked well for both the company and its monthly employees over the preceding four years so the board decided that "approximately six percent of the net income before bonus and income tax, be set aside and distributed to monthly or salaried employees of the Company."

The November 24 board meeting minutes were unprecedented in that, for the first time, data on the company's financial performance over the preceding four fiscal years was recorded as shown on the next page.

Company officers made a point to conserve finances through fiscal years 1951 and 1952, not knowing for certain how much the Korean War would affect production of commercial delivery trucks. Directors approved salary increases for salaried employees and a $5.00 per share stock dividend, payable in January 1952. The initial stock dividend approved for payment in January 1953 was only one dollar, indicative of the board's exercise of fiscal restraint in an unstable business environment. However, in July 1953, the directors sensed that the company's financial position was stable enough to proceed with a subsequent dividend payment of $4.00 per share.

This radically shortened circa 1950 Urban body is mounted on a shortened Chevrolet forward control chassis. A sliding side door wasn't feasible on such a short body and the customer opted for a special piano hinged door in lieu of a standard production jackknife door. It's presumed that this truck had additional running gear enabling it to operate on railroad tracks as a track inspection vehicle.

The bonus plan that was instituted for salaried employees in 1947 took a new twist at a September 28, 1953 board of directors meeting when the board resolved, "that in order to provide the salaried employees of the company with a program which will assist them in meeting with the contingencies of death, disability, loss of employment and old age dependency, we hereby adopt a plan to be known as "The Profit-Sharing Retirement Plan". Whether this plan was compassionate, paternalistic or shrewd depended on one's point of view. Monetary bonus payouts were stopped and replaced with whole life insurance policies on each salaried employee. As the modest whole life policies built cash value and paid dividends, company officers directed the insurance company to apply the dividends toward the cost of annual insurance premiums, gradually reducing and eventually eliminating the annual life insurance premiums paid by the company. If the employee left the company, the policy was cancelled and the accumulated cash value of the policy was returned to the company coffers. An additional benefit to the company was that the insurance premiums were deductible items for federal income tax purposes.

Fiscal Year Ending	Net Income Before Bonus	Bonus Payout	Date of Payment
Sept. 30, 1947	$315,266.68	$3,600.00	Dec., 1946
		$7,700.00	April, 1947
		$9,250.00	Sept., 1947
Sept. 30, 1948	$402,266.68	$3,710.00	Dec., 1947
		$14,450.00	April, 1947
		$11,825.00	Sept., 1948
Sept. 30, 1949	$160,193.55	$3,780.00	Dec., 1948
		$3,366.50	Sept., 1949
Sept. 30, 1950	$109,798.20	$1,700.00	Dec., 1949
		$1,900.00	June, 1950
		$2,175.00	Sept., 1950
	$987,918.54	$63,456.50	

Production was stymied in mid-1952 by a steel strike that shut off the flow of incoming chassis from truck manufacturers as well as the sheet steel that Montpelier needed to fabricate truck bodies. Montpelier kept up its normal manufacturing pace with its on-hand inventories but by late July the chassis inventory was exhausted and sheet steel was virtually impossible to find. Harry Schwartz explained that the company was "living off its fat" during the strike and ultimately had no choice but to lay off workers and inform customers that truck deliveries would be delayed. The strike was resolved in August and by early September Montpelier Manufacturing's recovery phase was well along. A slumping economy had prompted Montpelier's customers to postpone orders earlier in the year but once the steel strike was settled, customers decided to place orders before the supply of trucks was stifled once again due to some other unforeseen event.

Montpelier Manufacturing's 25th anniversary was celebrated in December 1952 (actually eight months after the milestone was achieved) with an elaborately planned dinner for 350 employees and their families held at the Montpelier High School gymnasium. Bill Sprankell, Dick Parker and Ferman Haase each received gold watches commemorating 25 years of service with the company. Harry Schwartz, whose official title at the time was Secretary-Treasurer of Montpelier Manufacturing, presented the watches. The three senior employees also received a gold fountain pen and pencil set, as did Carl Henry, William Steinke, Marvin Hiner, Wayne Bordner, Albert Meiser, Verlyn Heller, Doran Bavin and Bill Freese, employees who had completed 15 to 24 years of service. Employees with ten to 15 years of service each received engraved pencils.

In a show of appreciation for Harry Schwartz's faithful leadership through the many incredible challenges the company faced in the years leading up to its silver anniversary, Montpelier Manufacturing employees gave Harry a commemorative plaque on which was engraved the image of a 1952 Urban delivery truck and one of Montpelier's earliest trucks.

9
Building Truck Bodies, Montpelier Style

Hubert Hutchinson and Keith Miller provide an insight into post-war life on Montpelier's shop floor. Hubert resumed civilian life in January 1945 as a gas station attendant earning one hundred dollars a month. Three months into the job, Bill Steinke and Ferman Haase pulled into the station and offered Hubert a job at Montpelier Manufacturing, starting the very next day. Hubert accepted and started working for Steinke in the pattern shop for fifty cents an hour. Six weeks later, he seized an opportunity to work in the body welding and assembly department for Bill Freese, a department that he remained in during the rest of his tenure at Montpelier Manufacturing. As good now as he ever was with figures, Hubert still recalls the substantial pay increase that came with his first and only job transfer within the plant. Moving from a job that paid a low hourly wage to one that paid on a piece rate basis enabled him to earn a respectable living by hanging side and rear doors. Hanging a pair of sliding

A top jig used to check for proper fit of the six to eight individual curved steel panels that comprise a complete top. Two electric motor driven forming hammers can be seen in the background along with a few very complex hammer formed panels at the far left. The power hammers were used to form complex sheet metal conturs for fenders, front end panels and the four corner panels on each top.

side doors paid $2.20, a single rear door was worth $1.50 in wages, a pair of narrow rear doors paid $2.25 while a pair of wide rear doors earned Hubert $4.50. Jackknife side doors were more complicated to hang but paid a handsome wage of $8.00 per pair. Hanging sliding or jackknife side doors included the precise task of aligning and welding the door track into place at the top and bottom of the door aperture. Hinges had to be installed on the body and rear doors so that the doors opened and closed without binding and door latches were properly aligned with the doors closed.

Max Kunkle made side and rear doors in the welding department located at what was called the "back end" of the plant. Doors consisted of plywood cores with steel interior and exterior skins pressed over the plywood. After the steel panels were formed around the plywood core, the joints between the interior and exterior door skins were soldered to prevent water intrusion. Hubert's tool of choice for hanging rear doors was a five-pound sledgehammer. It was rare when a delivery body's rear door aperture was square and plumb when it left the welding department. Hubert's judicious hammer blows achieved the required improvements and enabled him to properly fit and install rear doors.

The dimensional accuracy of doors sometimes varied from batch to batch due to mishandling of the jigs that were used to assemble the doors and the fact that the jigs were stored outside and unprotected from the elements. Hubert had to be prepared for a variety of potential problems with each pair of doors he hung. He often wondered how replacement doors were installed on damaged customer trucks when one of the more talented and versatile body men in the "catch up department" was dispatched to the foreign soil of a customer worksite in Chicago, Philadelphia or New York to make on-the-spot field repairs.

When there were no more doors to be hung, Hubert installed plywood paneling in Urban Model 9 or 11 bodies for a piece rate of $3.50 per truck or he installed insulation between inner and outer truck body panels for $3 to $5 per truck, depending on truck size. Interior panels were nailed to the wood framing of a Model 9 or Model 11 body and an aluminum molding strip was inserted between the edges of two adjacent plywood panels to cover the nails and the small gap between the panels. Steel exterior panels were fastened to the wood framing with drive screws (nails with a helical or twisted shank that threaded itself into the framing as it penetrated the wood) that were hammered in place. A final one or two hammer blows partially embedded the drive screw heads into the surface of the steel panel. The drive screw heads were then "buzzed" with a grinding disc so that only the tapered countersink portion of the heads were left to retain the steel panel. Leo Mansfield or Keith Miller then came along to spread solder to cover the seam and the adjacent drive screw heads.

A layoff loomed whenever there wasn't enough work to keep Hubert busy. Fortunately, Hubert's foreman, Bill Freese, appreciated Hubert's valuable skills as a door hanger and welder and made sure that he was rarely out of work.

Hubert worked at Montpelier Manufacturing until 1955 when he was hired by the Ohio Turnpike Commission as a service station manager. He returned to his old job at the body works two years later, staying until 1962 when he started working full time as an independent insurance agent.

Hubert recalls that "Every time you did something you had to write it down. At the end of the day you'd make out your pink tickets. You'd put all of what you did on them and add it up and see how much you made. Of course, I was always three, four or five days ahead of time. I could have gone in for three, four or five days and not done a thing and still made $20 a day.

"When trucks were selling well, there was plenty of work to do. Then there would come a lull time. During December, January and February it was slow. They would just lay you off in the slow months. They usually kept me so that they had me around to do some of this *(miscellaneous)* stuff. If you didn't have piecework to do, they'd pay you by the hour for doing other stuff but it was only $1.50 an hour and your paycheck didn't look too big.

"We had a guy by the name of Ford Hollstein that, according to his foreman, was the best guy we ever had working for us. He never stood around with his hands in his pockets because his arms are too short t reach his pockets!" The short reach of Ford's arms was the primary cause of an unfortunate injury that is described in the pages that follow.

Keith Miller also started at the body works in 1945. If bodies suffered small dents or dings as they were assembled and welded, it was Keith's job to dolly the dents out before the bodies proceeded to the paint room.

He also soldered seams on bodies and tops, a skill that he could perform fifty percent faster than others doing similar work due to previous experience in auto body repair. Unfortunately, his ability to solder seams at a quick pace didn't translate into increased pay because of limitations on the amount of piece rate work he and others were allowed to turn in for a day's work. A multi-skilled individual, Keith recalls that "Some of the guys knew what they were going to do when they came to work in the morning, but I didn't because I was doing everything."

When grinding solder, occasionally a grinding disc would suddenly become snagged on a rough spot just below the layer of solder and wrench the grinder away from an unsuspecting operator's hands. On one occasion, a wayward grinder tore off one of Keith's pant legs, fortunately avoiding any contact with skin, flesh and bone. Others grinding solder joints weren't as fortunate. When band aids, mercurochrome and cotton swabs obtained from the stockroom's first aid kit weren't up to the task of dressing an injury, a wounded worker's foreman usually whisked him off to a doctor's office or the local hospital emergency room for stitches and then taxied him home for the remainder of the day. When grinders weren't cutting flesh and skin, they would occasionally fling metal shavings into a worker's eye. (Obviously, safety glasses weren't a mandatory safety item on the shop floor.) The injured worker would make a quick sprint to the office in search of a volunteer willing to extract the offending particle. On occasions when successful results weren't achieved in the office, another trip for professional medial assistance was called for.

This view of Montpelier's press room shows two press brakes that were used to form a straight and relatively sharp bend in the sheet metal parts. Flat sheet metal blanks are stacked on the workbenches prior to forming. Master sheet metal patterns hang from the rafter at the extreme left. The workbench in front of the rotary shear was used to support a large sheet metal panel as workers fed the panels into the shear. The rotaryshear was typically used to cut irregular contours on certain steel blanks before the panel proceeded to one of the press brakes for the forming bends.

Several finished steel tops await transfer to the body assembly area to their immediate left. The six to eight individual panels that comprise a top can be seen in the top situated in the foreground. Tack welded joints between the individual top panels have been soldered and sanded.

On one occasion, Keith was finish grinding a truck body's windshield header while standing on the dashboard and straddling the windshield divider. The grinder's power chord came loose, electrifying the truck body and shocking poor Keith rather forcefully. A few excruciating seconds seemed like hours before Keith was finally able to pitch the disc grinder in whatever direction he could. He spent the next few minutes slumped over the windshield header, slowly gathering his wits and carefully checking to see if he was once again fully ambulatory.

Many of Montpelier's shop employees came from a farming background and, if nothing else, brought reasonably good common sense and nominal manual labor skills with them upon hiring in at the body works. However, pride in their straight rows of corn and soybeans didn't extend into their farm workshops which were often so cluttered that performing repairs in them were next to impossible. The cluttered workshop atmosphere at home seemed to carry over quite naturally on Montpelier Manufacturing's shop floor. No initiative was taken to make process or housekeeping improvements simply because of the habits many of the workers brought from their homes or farms and because they were compensated only for making production parts and performing production work, not maintaining a department in an orderly fashion. The disarray within the plant often forced shop workers to go and hunt for the parts they needed, an activity they were compelled to do because it was driven by an urgent need. The longer term benefits of keeping an orderly shop were just not that obvious to workers in their daily search for a part or tool to fill an immediate requirement.

Keith Miller's work habits contrasted with those of his fellow workers and prompted him to conclude that "They could have made half-again as much as they were putting out if they would have had it organized. Departments in the back end of the shop weren't fabricating parts in time for those who needed them

The power shear visible in the center of this photo was used to swiftly perform straight-line shearing operations on thinner steel sheets as well as thick plates. A punch press, visible at the extreme right, was used to punch holes or notches in sheet steel blanks or for more complex forming operations not feasible on Montpelier's press brakes.

to assemble bodies. Production wasn't well coordinated among the departments. I was surprised that they *(the trucks)* turned out as good as they did."

Keith had nearly made up his mind to leave Montpelier Manufacturing to start his own auto body repair shop on a full time basis when Bill Freese made him an offer that clinched the decision. According to Keith, "Bill Freese told me that there were a bunch of bodies that they brought in that needed repair and that they were going to pay me by the hour and the hourly rate back then was hardly worth anything." Keith found the offer unacceptable, almost to the point of being insulting, and replied, "You know, Bill, I'm going home, and don't call me." That was the occasion of his abrupt departure from Montpelier Manufacturing and going into his auto body repair business on a full time basis. The bodies needing repair were customer trucks that had been returned to the plant for repairs. Keith knew that Freese didn't have any room to negotiate a more favorable arrangement, fully aware that Harry Schwartz had given Freese strict marching orders. With his body repair business well established on a part time basis, Keith knew that he could make it a full time occupation and waved to his coworkers as he walked out the door.

Keith's brother Denver began his career at Montpelier Manufacturing immediately after graduating from high school in 1941. He was hired by pattern shop foreman Bill Steinke who had known Denver all through his years in school. An opportunity to work in the pattern shop was conferred on a select few employees and it was Denver's good fortune to be chosen to start working there. The pattern shop's primary function was to build prototype bodies or new derivatives of existing bodies for customer approval and to sort out potential production problems in a controlled, low volume laboratory environment prior to releasing a new or modified design to the shop floor for higher volume production. Under Bill Steinke's tutelage, Denver learned every skill that was involved in the manufacture of truck bodies.

The Montpelier Manufacturing engineering office, located in the basement below the rest of the office staff. A blueprint machine is visible against the wall at the far left. The flat drawers on the file cabinet in the far corner stored engineering drawings, allowing them to lay compactly in flat stacks.

By 1955, Denver was promoted to the position of foreman over the priming, painting, mounting, wiring and finishing departments, areas that workers referred to collectively as the "front end" of the shop floor. During his nine-year stint as foreman, he reported either to the plant superintendent or directly to Harry Schwartz. In early 1964, he somewhat reluctantly took on the role of plant superintendent, a pressure-cooker job that had gotten the best of several men in recent years. As superintendent, Denver's responsibilities encompassed all facets of production at a time when the plant was operating at breakneck speed. While striving to meet unprecedented demand for the profitable line of Montpelier Urbans and Side-Aisle Delivery trucks, Denver had the unenviable task of insuring that overbearing demand for the full line of unprofitable PDQ trucks was satisfied. Denver grappled with the demands of the superintendent's job until the end of his association with the Montpelier Division in mid-1965.

Denver's most vivid recollections of his career at Montpelier Manufacturing are tied to Harry Schwartz. Harry was involved in a minor auto accident while on business in Cleveland one summer at the same time that Denver was on vacation. Not wanting his damaged Thunderbird to be seen on the streets of Montpelier, he called Denver and told him that he'd be bringing the car out to Denver's home for immediate repairs. When the boss presented "opportunities" of this nature, they were virtually impossible to turn down but Denver declined, indicating that he was way behind on seasonal work that had to be done on his farm. Harry understood Denver's predicament and alleviated his concerns by generously offering to send out a few "helpers" from the factory to speed up the repair process. Realizing that there was no way of shirking the task, Denver agreed, fully aware that the so-called helpers would contribute nothing to fixing the problem at hand. Within minutes, the car and Denver's helpers were in Denver's barn. Denver grudgingly applied his auto body repair skills while his two assistants rested comfortably and observed his every move.

Each Montpelier Manufacturing salaried employee received periodic bonus payments in accordance with Harry's Schwartz's estimation of their value to the company. In 1953 Schwartz announced that he was replacing the bonus plan with a pension fund, a move that Denver immediately interpreted as a cost saving measure. One by one, Schwartz explained the plan to his salaried employees. He pitched the proposal to Denver, explaining that "Salaried employees would pay in so much and I would contribute so much and we'll do away with the bonuses. Do you want to do that, Denver?" Denver flatly declined the offer, indicating that he had faithfully taken the bonuses in the past and invested them for his own retirement security. Schwartz responded, "Yes, you might do that, but how about the other guys?" Denver replied, "I'm not worryin' about those other guys, I'm worryin' about myself." Schwartz replied, "Does that mean that you want my offer pushed right off the table?"

"No." Denver responded. "If you're not going to give me a bonus, then I'll take the pension."

Denver knew all along that the only reason Schwartz offered the pension plan was because he had something to gain by it. The pension offer was actually a life insurance policy that built cash value by paying dividends. The dividends would eventually become large enough to pay for the life insurance premium, freeing Schwartz and Montpelier Manufacturing from any financial obligation. If an employee left the company after enrolling in the pension plan, he only received the money he had invested in the plan. Any funds remaining in the employee's account were used by the company to pay its portion of the premiums for remaining employees.

As salaried employees, foremen and superintendents were required to attend Saturday morning meetings to plan work for the following week. Denver was supposed to show up for one of the regular Saturday morning meetings but he had a lot of to work at the auto body repair shop that he operated on his farm. He was held in reasonably high esteem by Harry Schwartz and could skip Saturday meetings occasionally without fear of reprisal if he was too busy at home. On this particular Saturday, Schwartz noted his absence, gave him a call and told him to come in. Upon arriving, Harry informed Denver that he had just bought a farm and that he wanted "someone to look after it, get the crops taken care of."

Like other foremen, Denver was a salaried employee and wouldn't receive extra compensation for being singled out by his boss for this special assignment. Instead, he would have to rely on Harry to give him money "under the table" as he frequently did for other employees that performed special assignments for him. Denver was overloaded with work on his own farm and his body shop, not to mention his day job at Montpelier Manufacturing, but grudgingly said yes to Harry's entreaty, realizing that agreeing to do the work was what Harry expected of him. It was a long harvest season and Harry kept after Denver until the final few acres of soybeans were harvested. Fortunately for Denver, Schwartz rented the ground to another farmer in succeeding years and Denver effectively "wiggled off the hook" as Schwartz's on-call farmer. Denver was fairly compensated for harvesting the crops but would have preferred to have spent the time pursuing his own endeavors.

In fairness to Schwartz, Denver is quick to say that Harry paid him well and treated him with a high level of respect.

In late 1953, Montpelier's board of directors altered the titles bestowed on two of the three company officers. Listed on the next page are the names of the company officers and their associated offices before and after the change.

No explanation for this change was given in board meeting minutes. However, insight into the reasons for the titular changes of 1953 is provided in the pages that follow. In 1957, Harry Schwartz's title was expanded to that of President, Secretary, Treasurer and General Manager.

The following year, Montpelier's board announced the appointment of John C. Rundell to the position of Vice President in Charge of Sales, "until further action by the Board." Barring unforeseen circumstances Rundell would be *appointed* each year by the directors, in contrast to the annual *election* of other higher-ranking company officers by the board. Rundell succeeded Gene Robertson, a former Chevrolet dealer from Crane, Missouri, who had joined Montpelier Manufacturing shortly after World War II as the company's first employee specifically hired to organize and direct the company's sales activities.

	Official Title, 1939-1953	Official Title, 1953 through 1957
Harry A. Schwartz	*Secretary, Treasurer & General Manager*	*President*
Grace L. Hill	*President*	*Chairman & Vice President*
Margaret H. Schwartz	*Vice President*	*Vice President*

The office where sales, purchasing and the clerical staff were domcicled. Harry Schwartz's office, furnished with large leather chairs, is visible through the open door at the far end. A switchboard is located on the desk just outside his office, close to the office entrance. A hallway just outside the entrance to the office acted as a buffer between the office and shop and served as a waiting room for visitors and job applicants.

10

Plastic Era

The first of three major product development programs embarked upon by Montpelier in the 1950s was the development of fiberglass reinforced plastic (FRP). With no immediate commercial application in mind, Montpelier Manufacturing started to experiment with plastic body panels in the early 1950's. Fiberglass reinforced plastic was more resistant to minor scrapes and dings than conventional steel or aluminum panels. Repairs were made with relative ease without having to remove the damaged panel from the truck. The highly skilled cutting, welding, hammering and soldering procedures required in repairing damaged sheet metal panels were replaced by repair techniques that were much easier to master. FRP offered greater corrosion resistance than steel or aluminum, an important consideration in humid saltwater coastal areas and northern regions where corrosive salts were applied to roads during the winter months. FRP was compatible with the same primers and paints used for steel and aluminum panels and had good insulating qualities, shielding cargo from temperature extremes and reducing the damaging affects of condensation in humid or cold climates. Fiberglass panels also resulted in a truck of lower net weight, resulting in

The photograph and accompanying text point out the virtues of Montpelier's new fiberglass reinforced plastic Side-Aisle Delivery bodies. The body is mated to a 1954 Ford cowl and chassis. Montpelier first introduced fiberglass reinforced plastic bodies as an alternative to conventional steel bodies in 1956, following a three year development program.

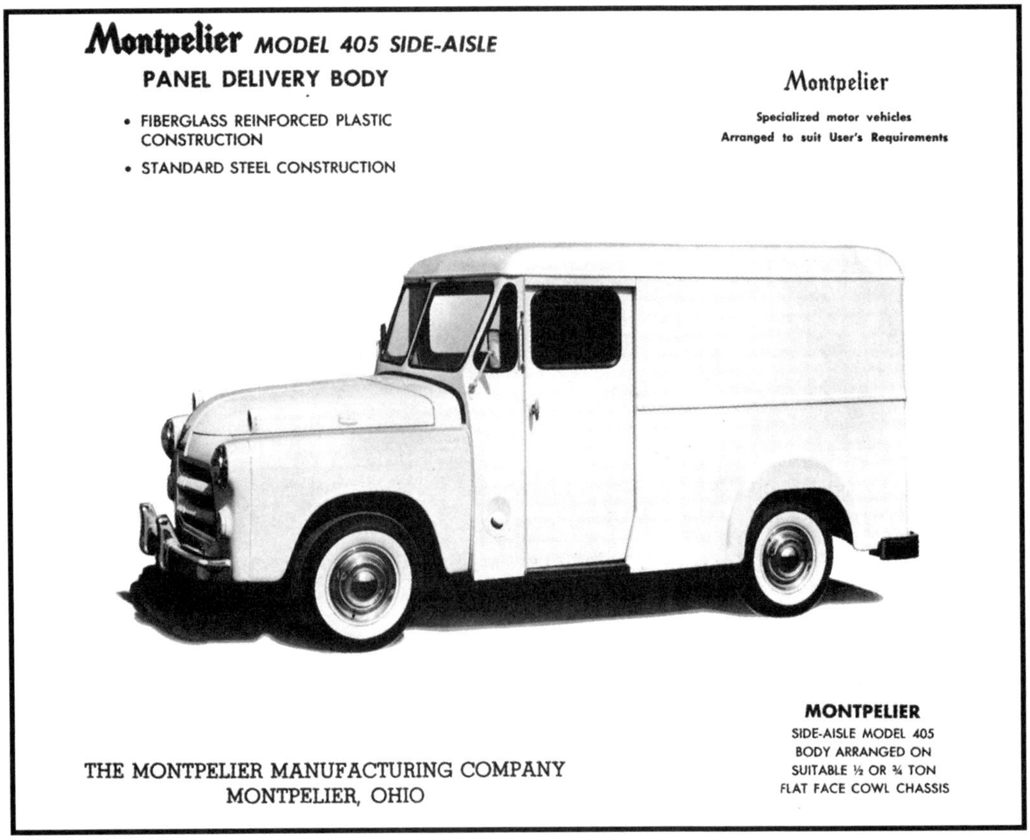

A Montpelier FRP Side-Aisle body mounted to a 1955 Dodge 1/2-ton cowl and chassis. The gas filler location is somewhat odd and differs sharply from the 1954 Ford pictured on the first page of the chapter.

reduced fuel consumption or enabling more of the truck's gross combined weight capacity to be devoted to carrying cargo instead of a heavier steel body. All fiberglass body panels produced for Montpelier delivery bodies were translucent in their pre-painted state.

The May 9, 1957 edition of Montpelier's *Leader-Enterprise* newspaper reported that a major breakthrough in the sale of trucks incorporating FRP panels occurred with Montpelier Manufacturing's receipt of an order for 165 trucks from American Railway Express. This was the first order that this prestigious customer had placed with Montpelier Manufacturing, an exciting sales conquest for the Montpelier firm. Railway Express opted for steel as the primary material for their truck bodies but roof panels were specified in the FRP material with portions of the roofs left unpainted. The unpainted translucent sections of a FRP roof allowed natural light to illuminate the cargo compartment, thus making it easier for route drivers to locate and keep track of their deliveries. The May 9 *Leader- Enterprise* story reminded readers that, "Montpelier Manufacturing originated the forward control truck body in the '30's, building the first prototype for International. Another innovation by the company was making the driver's floor at curb level, eliminating steps."

Although receiving the Railway Express contract was a major sales accomplishment and the occasion was used to loudly trumpet the use of FRP in body construction, it was by no means Montpelier Manufacturing's first foray into the sale of trucks with FRP bodies. Montpelier's early 1950's commitment to manufacturing FRP delivery truck bodies resulted in steady improvements to the molding process in the years prior to the prestigious Railway Express order. Achieving consistency and control over the process enabled them to start offering complete FRP truck bodies as an alternative to their established line of all steel bodies for the 1956 model year. By the time they had received the Railway Express order, the company reported that it had delivered several hundred FRP-bodied trucks that were proving themselves as a worthy alternative to steel-bodied trucks in daily delivery service. Targeted markets for Montpelier's FRP bodies included bakery, linen, laundry, dairy and food product deliveries where maintaining the desired appearance of the product and protecting the products from various environmental conditions were major concerns. It's likely that the studio type photos of FRP bodied trucks were passed out at trade shows and included in mailings to many prospective customers.

The May 9 *Leader-Enterprise* article also noted that Montpelier's work on FRP body panels first started in 1953 and that General Motors had presented the company with the opportunity of making 300 FRP Corvette bodies in June of that year. The mid-1953 General Motors opportunity seems hard to reconcile with Montpelier's brief and limited experience in the production of FRP body panels and the fact that they were still two years or more away from offering FRP as an option for their own truck bodies. Montpelier passed on the GM offer, dissuaded by the large tooling investment and the possibility that GM would eventually want to work with its Fisher Body Division once production volumes became significant.

The hiring of Harold "Buzz" Bartick in July of 1954 was instrumental in accelerating the development of Montpelier's line of FRP truck bodies. Bartick came to Montpelier from the Crosby Aeromarine Company, a division of Midwestern Industries, Inc. that manufactured FRP boats in Harlan, Indiana, a small town about 40 miles southwest of Montpelier. Among those working with Bartick in Montpelier's plastic molding department were Kenneth Harman, Jerry Kissinger and Paul Strawser.

Even though John Rundell had primary responsibility for Montpelier Manufacturing's sales efforts, serving first as Fleet Sales Manager and then as Vice President in Charge of Sales, Harry Schwartz was credited with closing the deal on the new contract with Railway Express

Robert C. "Jumbo" Kunkle served the company under John Rundell as Montpelier Manufacturing's fleet sales manager. "Jum," as he was known by his fellow workers, was charged with the task of building on earlier sales successes and Schwartz's Railway Express sales conquest to convince existing and prospective fleet customers of FRP's many advantages. It was hoped that Sears Roebuck, Marshall Fields and Omar Bakery Company, Montpelier's largest fleet customers at the time, would recognize the many good features of FRP and opt for it on future truck orders. In the early 1950's, Jum and fellow sales associate Bill Sayers were on the road visiting customers, very often on a weekly basis, maintaining close customer contacts and working intently on steering orders toward Montpelier and away from competitors. Bill left Montpelier Manufacturing for a fleet sales manager's position at one of Montpelier's competitors, the DeKalb Commercial Body Company of DeKalb, Illinois, in 1953 but years later would return to Montpelier Manufacturing in a completely different capacity.

John Rundell, Montpelier Manufacturing's Vice-President in Charge of Sales, inspects the pristine interior of a Montpelier Side-Aisle Delivery truck with body panels made entirely of fiber-glass reinforced polyester plastic. The truck was on display in the Montpelier Manufacturing booth of a national trade show, exact location and date is unknown.

Jum began his sales career at Hawk's Men's Store in Bryan, Ohio immediately after graduating from high school. Even as a teenager, he possessed and an innate ability to hook a customer. Merely by setting up a display of some of Hawk's less desirable ties in the store's basement, Jum was able to dispose of most all of the drab and unsightly cravats without resorting to a discount. The perceived discount of basement merchandise was all that was needed to prompt customers into making a purchase. Jum was an easygoing, generously proportioned and affable fellow, characteristics that would serve him well for nearly ten years in the sales department at Montpelier Manufacturing. Given Jum's natural salesmanship abilities and the many advantages FRP bodies offered, it was felt that sales of FRP bodied trucks would play a significant role in securing Montpelier Manufacturing's future.

With the introduction of its 1957 model lineup, Montpelier was producing seven Urban Low-Step Delivery models of all steel construction with cargo compartment sizes ranging from 250 to 555 cubic feet. Four of the seven steel-bodied models had identically sized counterparts in the FRP body lineup with cargo compartment sizes in the 250 to 445 cubic foot range. Steel bodied Side-Aisle Delivery models were offered in four sizes and for each body size there was a direct FRP-bodied equivalent.

Montpelier's data sheets with specifications, ordering procedures and pricing information for FRP Urban and Side Aisle Delivery bodies, published in September 1956, are reproduced below. FRP bodies by themselves were 30 to 40 percent more costly than steel bodies of equivalent size but the sharp differential in body cost was not as obvious when the cost of the truck chassis and the remaining components of a completed delivery truck were factored in. The cost of Montpelier's hybrid truck bodies, i.e., those having a fiberglass top in conjunction with steel side and rear body panels, undoubtedly fell somewhere in between the full steel and all fiberglass construction.

Molds were available for a full range of Urban and side-aisle top panels, side panels, doors and rear sections. FRP rear fenders were produced for some of the larger Urban models, other bodies sported FRP hoods and, over time, Urban grille bars were converted to the FRP material. With production of the newer fiberglass and steel bodied Urban models well under way, Chuck Brigle, one of the few employees that could see past Harry Schwartz's blusterous personality and appreciate his depth of character and commitment, decided to leave after fourteen years of service to the company. In Chuck's words, "That was a day that I dreaded. I went in and told him that I was going to leave him and go to the Aro *(a larger manufacturing firm in Bryan, Ohio)*. I was standing in front of him at his desk and *(in characteristic Schwartz fashion)* he shot up out of his chair and took me by the arm and threw me into one of his big leather chairs and said, 'You sit right there!' He went over and closed the door to his office and went back to his desk and sat for awhile. Then he got up and sat in the other leather chair beside me and he said 'After all these years, you

Prices, Specifications, Data and Ordering Procedures for Montpelier Low-Step Delivery Bodies

Model Number	Length on Floor	Inside Width	Inside Center Height	Load Space, Cu. Ft.	Wheelbase	List Price	
8	96	68	68	250	104	$1,520	Forward control 3/4 to 1 ton chassis, furnished by dealer
8810	120	68	68	315	116 to 125	$1,580	Forward control 3/4 to 1½ ton chassis, furnished by dealer
10	120	75	72	370	116 to 125	$1,600	Forward control 3/4 to 1½ ton chassis, furnished by dealer
101	120	75	76	390	116 to 125	$1,970	Forward control 1 to 1½ ton chassis, furnished by dealer
12	144	75	72	445	129 to 137	$1,730	Forward control 1 to 1½ ton chassis, furnished by dealer
121	144	75	76	470	129 to 137	$2,080	Forward control 1 to 1½ ton chassis, furnished by dealer
15	180	75	72	555	153 to 158	$2,270	Forward control 1½ to 2 ton chassis, furnished by dealer

Effective September 25, 1956

Chassis rquired for the Urban Low-Step Delivery Bodies are the forward control models - stripped type without sheet metal. Conventional chassis may also be used, providing they are altered by us for forward control operation. Bodies are also available for forward control chassis with front end sheet metal. All chassis supplied for the Urban Low-Step Bodies are to be complete with proper wheels and tires, side mount gasoline tank, suitable type front bumpers, head lamps, parking lamps, etc.

mean you're gong to leave me?' I said, 'That's my intention.' Harry replied, 'Aro can't pay you money like I do. If I'm not paying you enough right now, let's negotiate.' I said 'No, I don't want to negotiate, I just want a change. I want to do something different.' And he couldn't understand that. He gave me all the reasons why I should stay. He told me that Ferman *(Haase, superintendent of the pattern shop, sheet metal and welding departments)* was at the age where he was going to have to step down and that 'I've been grooming you for years.' 'I know you have, but I'd just like to try something different.' And he could not understand that but he never got angry. Finally I said, 'I'm still going to leave.' Harry said, 'Well, good luck. You come back in a year from now and tell me that you're not sorry that you left.' And I did. I went back. One day I dropped in again and I saw him. We had a nice chat and I told him what I was doing and using everything he taught me. I wasn't sorry that I left and I wanted him to know that. We were always good friends after that."

The Fiberglass Molding Process:

Each Montpelier fiberglass reinforced plastic body panel was produced one-at-a-time in one of Montpelier's many large open cavity molds. The molds were constructed of fiberglass and wood and were made by Montpelier Manufacturing employees under the guidance of Buzz Bartick. Limited floor space dictated that the molds be stored outside while not in use. "Stored" is a somewhat misleading description for the manner in which the molds were treated when not in use. Initially they were scattered on the facility's south lot, exposed to damage from the elements and routine traffic in the lot. As the number of molds proliferated, an open-sided metal shelter with vertical storage racks was built in 1963 to better protect the molds from damage and to reclaim space in the factory's south lot. The quickest and surest way to retrieve a mold from storage was for six or more men to carry it into the shop. Once indoors, the molds could be moved about on a cart that was pulled by a towmotor or forklift to the plastic molding department located on the south side of the plant, roughly two-thirds of the way down from the office building. After a mold was jockeyed into the plastics department, it was placed on the floor and thoroughly cleaned. Steel strips were installed roughly four inches in from the perimeter of the mold, the strips providing a smaller and precise sharp edged perimeter for the body panel that was to be molded.

Returning a mold to its storage location in the outdoor shelter was a physically demanding task as well. One occasion found Ford "Shorty" Hollstein helping five of his co-workers store a large mold used to make FRP tops for Railway Express trucks. While the other men held the mold at chest height, Shorty had to fully extend his arms at the end of the mold as it was coerced into its deep horizontal storage rack. Shorty failed to remove a hand as the mold abruptly slid into position and, as a result, lost a finger as it was sheared off by the heavy mold and the storage shelf.

Prices, Specifications, Data and Ordering Procedures for Montpelier Side-Aisle Delivery Bodies

Model Number	Length on Floor	Inside Width	Inside Center Height	Load Space, Cu. Ft.	Wheelbase	List Price	
409	79	63	57	160	108 to 115	$1,045	1/2 or 3/4 ton chassis, furnished by dealer
419	92	63	57	185	116 to 123¼	$1,120	3/4 to 1½ ton chassis, furnished by dealer
415	100	68	63	240	116 to 123¼	$1,230	3/4 to 1½ ton chassis, furnished by dealer
420	114	68	63	275	126 to 135	$1,400	1 to 1½ ton chassis, furnished by dealer

Effective September 25, 1956

Chassis for Side-Aisle Delivery Bodies are flat face cowl chassis, complete with hood, cowl, radiator shell, front bumper, proper tires and wheels, front fenders, side mount gasoline tank, tire carrier, etc.

The first step in the molding process involved spraying the mold cavity with a wax release agent. Polyester resin was dispensed from a 55-gallon drum as a viscous liquid the consistency of cake batter into metal or cardboard cans scrounged by the plastics department. One of the more reliable sources of cans for this purpose were discarded cardboard motor oil cans that plastic department workers routinely picked up at local gas stations. Coffee cans were also used whenever they could be found. A catalyst that served as a hardening agent was added to the polyester resin and mixed in the oil or coffee can with a mixer blade driven by an electric drill. As soon as mixing was completed, the liquid resin was poured into the mold cavity and spread out over the entire mold surface with rubber paddles and rollers, making sure that the thickness of this initial layer was adequate and that any entrained air bubbles were purged. A large piece of woven fiberglass cloth, previously cut to the correct size and shape, was spread out over the initial layer of resin as it began to harden. Wrinkles in the fiberglass cloth were smoothed out and another can of the resin and hardener mixture was poured over the cloth. Squeegees were used to spread this second application of resin and incorporate it into the all of the minute voids in the loosely woven fiberglass cloth. A one-eighth inch thick fiberglass mat was then laid in the mold and smoothed out. With considerable force, rubber rollers were used to roll in a third and final application of resin and hardener, compressing the fiberglass mat while insuring that this final application of resin thoroughly penetrated the voids in the fiberglass mat with enough resin remaining on the surface to conceal the glass fibers and provide a smooth surface in the fully hardened condition.

If the panel being molded was a fiberglass top, a seasoned oak framework was embedded in the third application of resin before it hardened. Wooden reinforcing ribs went across the tops and two longitudinal reinforcing ribs ran the length of the top, each about one-third of the way in from the top's longer sides. All of the reinforcing ribs interlocked with wooden mounting rails that formed the top's perimeter mounting flange. Another sheet of fiberglass cloth was carefully laid in over the wood frame and a fourth application of resin was applied and worked into the cloth, essentially embedding the oak frame within the layers of resin and fiberglass. The oak reinforcing frame was produced in Montpelier's mill room.

Once the molding process began, workers were in a race to add fiberglass cloth or matting and apply successive layers of resin. If the resin hardened in certain areas before it could be absorbed by the fiberglass cloth or mat, these areas had to be cut out and repaired by repeating the entire process in the defective area with considerable overlap into the solid areas. The sooner a defect was identified and removed the better, as the tops were much more difficult to salvage once the resin reached full hardness. After the resin and fiberglass lay-up process was completed, the molded body panel remained in the mold for an additional half-hour until it was fully hardened. A four man team was required to mold a large Urban FRP top panel, taking nearly one and a half-hours to complete the job. Smaller FRP panels took just as long to mold but required less manpower.

Adding the catalyst or hardening agent to the polyester resin generated quite a bit of heat, which is what initiated and sustained the hardening process. If the paint rapidly burned off the outside of a coffee can shortly after the resin and hardener were mixed, it was a sure sign that an excess amount of hardener was used and that the mixture could not be spread over the entire mold surface before it solidified. If too much hardener were used or an excess amount of hardener and resin mixture were concentrated in one area of the mold before it was spread out over the entire mold cavity, the area of concentrated heat could cause the spontaneous ignition and a rapid response from the fire brigade. More of the hardening agent was used during the winter months to compensate for the cooler temperatures in the plant.

The polyester resin was often splattered on the floor as it was poured and distributed in the molds. Cardboard was hastily laid on the floor over spilled resin so that the somewhat hectic pace of work could proceed without interruption.. Montpelier's plastic molders were easily distinguished from other shop employees by the layers of resin that had accumulated on their work shoes.

After the team of molders extracted a molded top, rear section or side panel from its mold, the freshly molded panel was transferred to one of several finishing stations that occupied the remaining half of the plastics department. With the molded panel resting on several bucks, one or two workers sanded the exterior surface of each panel with electric grinders fitted with seven-inch sanding disks. Entrapped air voids and smaller areas with defects that weren't detectable while the panel was still in the mold were repaired, most often along the edges of the panels where it was difficult to force the liquid resin to fully conform to the mold's sharp corners. The repairs were accomplished by mixing a powdered material, resin and hardener into a putty-like substance, filling the voids and sanding them smooth after the mixture hardened.

The sanding process generated large quantities of dust and debris that irritated the skin. A hood directly over the sanding station evacuated most of the dust generated during sanding. However, debris came up directly into the face of the person doing the sanding. Breathing masks were provided to keep the dust and debris out of the workers lungs. Workers wore the masks if they cared to, but safety glasses hadn't yet worked their way into regular use in the Montpelier plant. No respirators were used by the molders for protection from fumes generated during the molding process. Acetone, a volatile and highly flammable liquid subject to strict environmental limitations today, was used to clean rollers, squeegees and other tools used to spread the polyester resin during the molding process.

There were no safety showers in the plant to resort to in the event of a chemical burn or other accident. A sink was located in the plastic department where employees could wash their hands and arms. Workers were careful to rinse their forearms in a stream of cold water to wash away the irritating fiberglass particles before they scrubbed their hands and forearms with soap. Cold water kept the skin's pores closed so that glass fibers were less likely to become embedded.

Bill Gipple, one of the more experienced workers in the plastics department, recalled that the hard working crews in the plastics department complained to Adam Alguire, their foreman, about the low hourly wage they received for the amount of product they made. Management responded by converting to a piece rate form of compensation on a trial basis. After realizing that management had set the rates at a level that had virtually no impact on their earnings, the effort was abandoned and the department reverted to hourly compensation.

Bill and two others in the plastics department took welding classes due to the ever-present need for good welders and the lure of greater pay. All three men became capable welders but to no avail. Instead, the company chose to hire new employees as welders, keeping Bill and his co-workers in the plastics department where their specialized skills could be put to the best possible use.

Three different molds were in use in the plastics department at the same time when the department was operating at full capacity, each mold requiring four workers during the lay-up process. Up to six additional workers would be sanding and finishing recently molded FRP panels while a few additional workers would be cutting fiberglass cloth and mats to the required shape in support of the molding operation.

Finished FRP body panels were stacked up against the wall in the southeast corner of the plastics department, ready about a week before they were needed by the body shop. Two shifts typically staffed the plastics department and a smaller third shift would be added during periods of peak demand. Initial demand for complete FRP bodies or an FRP top mounted to a steel body was minimal, eventually picking up to a point but never equaling the production rate of traditional all-steel bodies.

Two noteworthy employees in the plastics department were the Thompson brothers who had migrated to Montpelier from Tennessee. They not only performed respectably as plastics department workers, they were held in highest esteem by some Montpelier Manufacturing employees for returning from occasional trips to their native state with copious amounts of Tennessee moonshine.

FRP tops would often warp after molding, in spite of the stiffness provided by the top's wood framing. Body assemblers resorted to clamps to close the joint between the FRP top and the mating flanges of the body panels before installing screws through the body flanges and into the top's oak or steel frame.

The development of fiberglass reinforced plastic was the first of three major product development programs in the 1950's, all of which were resounding successes from an engineering standpoint. Collectively, they catapulted Montpelier Manufacturing into a uniquely prominent position among the company's many competitors in the delivery truck body building industry. Two of the programs, including the development of Fiberglass Reinforced Plastic, were commercial successes and were instrumental in gaining new business. A third program, described in Chapter 12, brought unforeseen complexities to the business.

II

New & Improved 1957 Urban Models:

Low-Step Urban bodies were fresh new designs for the 1957 model year, and marked the completion of the company's second significant engineering project of the 1950s. A total redesign of the Urban line of forward control delivery trucks was required in order to incorporate the newly developed FRP body panels. The new Urban bodies made it possible for customers to specify any combination of fully interchangeable steel or FRP body panels to meet their specific delivery or other vocational requirements. Low-Step Urbans also featured a new steel-fabricated front clip that would eventually be shared with a future product that was still in the development and testing stages.

Montpelier Manufacturing introduced a redesigned line of Low Step Urban models in the Spring of 1957. Two local businesses were first to take delivery of the new Urban models. The Urban purchased by Allen-Cunniff Furniture is shown here. The Montpelier City Laundry took a delivery of a similar model. *Photo courtesy of Allen-Cunniff Furniture, Inc., Montpelier, Ohio.*

First deliveries of the new Low-Step Urban trucks were made to Allen-Cunniff Furniture and City Laundry, two local Montpelier businesses. The occasion was marked by Montpelier's *Leader-Enterprise* newspaper with a photo showing John Rundell, Montpelier Manufacturing's Vice President in Charge of Sales, handing the keys for the first truck to Ed Cunniff. John Firestone was shown in a second photo, standing in front of his newly acquired City Laundry truck.

The forethought that went into the design of the Low-Step Urban bodies enabled Montpelier Manufacturing to introduce two Montpelier Super-Urban models. Super-Urbans were high-rise derivatives of Low-Step Urbans that offered up to seven hundred cubic feet of cargo volume, achieved primarily by a cargo compartment height of eighty-one inches. Aimed squarely at the parcel delivery market, Super-Urbans were enthusiastically welcomed into the extensive Railway Express fleet of parcel delivery trucks and into many smaller fleets operated by regional parcel delivery companies.

SPECIFICATIONS*

BODY DIMENSIONS

GENERAL DIMENSIONS IN INCHES

	REAR WHEEL HOUSE	NOMINAL INSIDE LENGTH (feet)	CAPACITY (cu. ft.)	INSIDE LENGTH	INSIDE WIDTH	INSIDE HEIGHT	APPROX. FLOOR HEIGHT LOADED	APPROX. OVERALL HEIGHT LOADED	OVERALL WIDTH	APPROX. OVERALL LENGTH	REAR DOOR WIDTH & HEIGHT	DESIRED WHEEL BASE
URBAN-VAN MODELS												
MCV-200-16W	YES	16′	875	187″	89″	88″	37″	127″	95½″	251″	80″ x 80″	129-133″
MCV-200-20W	YES	20′	1090	235″	89″	88″	37″	127″	95½″	299″	80″ x 80″	153-157″
MCV-200-24W	YES	24′	1300	281″	89″	88″	37″	127″	95½″	347″	80″ x 80″	171-175″
MCV-200-16	NO	16′	900	187″	89″	88″	43″	133″	95½″	251″	80″ x 80″	129-133″
MCV-200-20	NO	20′	1120	235″	89″	88″	43″	133″	95½″	299″	80″ x 80″	153-157″
MCV-200-24	NO	24′	1330	281″	89″	88″	43″	133″	95½″	347″	80″ x 80″	171-175″
URBAN-DELIVERY												
MUV-100-15W	YES	15′	800	180″	89″	88″	36-38″	127″	95½″	273″	80″ x 80″	153-154″
MUV-100-15	NO	15′	800	180″	89″	88″	42-44″	134″	95½″	273″	80″ x 80″	153-154″
SUPER-URBAN												
1625	YES	16′	700	192″	80″	81″	32-35″	118″	92″	293″	68″ x 73″	166″*
1225	YES	12′	525	144″	80″	81″	30-34″	116″	92″	245″	68″ x 73″	129-137″

*WHEELBASE CHANGE REQUIRED

STANDARD EQUIPMENT

	URBAN VAN	URBAN DELIVERY	SUPER URBAN
FRP ROOF WITH TRANSLUCENT LIGHT AREAS			■
STEEL ROOF WITH FRP TRANSLUCENT PANELS	■	■	
DUAL ELECTRIC WINDSHIELD WIPERS	■	■	■
REAR VIEW MIRRORS	■	■	■
FRONT VENTILATORS	■	■	■
AUXILIARY SEAT	■	■	
COACH TYPE CAB DOORS	■	■	
SLIDING CAB DOORS			■
LOAD SPACE LINING: PLYWOOD	■	■	
LOAD SPACE LINING: OFFSET STEEL			■
LOAD SPACE FLOOR: HARDWOOD	■	■	
LOAD SPACE FLOOR: TREADPLATE			■
SQUARED WHEEL HOUSINGS	■	■	■
REAR BUMPERS—RUBBER	■	■	
REAR BUMPERS—STEEL CHANNEL			■
REAR FENDERS	■	■	■
SKIRTLINE RUBRAILS	■	■	■
ROLL-UP REAR DOOR	■	■	■
DIRECTIONAL SIGNALS	■	■	■
CLEARANCE LIGHTS & REFLECTORS	■	■	■
DOMELIGHT—FRONT & BACK	■	■	■
WALK-THRU: CAB-TO-LOAD SPACE		■	■
PRIME COAT FINISH	■	■	■

*BLUE BLOCK INDICATES RECOMMENDED SPECIFICATION.
*ALL SPECIFICATIONS AND PRICES ARE SUBJECT TO CHANGE WITHOUT NOTICE.

OPTIONAL EQUIPMENT

	URBAN VAN	URBAN DELIVERY	SUPER URBAN
FORWARD CONTROL CHASSIS CONVERSION	■		
DOUBLE 40″ SIDE DOORS	■	■	
TAILGATE WITH DOUBLE REAR DOORS	■	■	
ORSCHELN HAND BRAKE ATTACHMENT	■	■	■
CAB HEATER AND DEFROSTER	■	■	■
DUAL SUN VISOR	■	■	■
FOLDING CAB DOOR		■	
COLOR FINISH—ONE SHADE	■	■	■

CHASSIS REQUIREMENTS

	URBAN VAN	URBAN DELIVERY	SUPER URBAN
CONVENTIONAL CHASSIS	■		
FORWARD CONTROL CHASSIS		■	■
GROSS VEHICLE WEIGHT (pounds)	18,000-20,000	15,000-18,000	10,000-18,000
FRONT AXLE (pounds)	6,000-7,000	5,000-7,000	4,500-5,000
FOUR-SPEED TRANSMISSION		■	■
AUTOMATIC TRANSMISSION	■		
POWER BRAKES	■	■	1625
POWER STEERING	■	■	

WRITE — WIRE or PHONE (5-3161) FOR FULL INFORMATION

THE MONTPELIER MANUFACTURING COMPANY, MONTPELIER, OHIO

A Model 1225 Super-Urban photographed in Montpelier before shipment to one of many REA terminals strung out across the contiguous United States.

Mid-50s Chevrolet hubcaps, a primed but unpainted body and a newer Urban front clip design provide clues that this truck was a prototype that Montpelier Manufacturing hadn't yet released for full production. The 1957 model year was the production debut for Urban deliveries very similar to this prototype.

Montpelier Manufacturing fabricated the roof of this work truck body and everything from the sliding side doors to the very rear of the truck. Ford supplied the windshield and front cowling as well as the forward control chassis. Nearly all of Montpelier's forward control delivery and work truck bodies were built on stripped chassis with front clips of their own design, making this Montpelier-Ford hybrid body a rare exception.

This photo of a Model 1225 Super Urban delivery body mated to a GMC 1½ ton chassis was taken at the Williams County Highway Garage, as were quite a few other photos of Montpelier bodied trucks. The swing-up awnings and the drop-down platform at running board level indicate that this truck was destined for a vocation that served walk-up customers.

Montpelier Manufacturing built six of these large Urban Delivery trucks and configured them as paddywagons for the Cuyahoga County (Ohio) Sheriff's Department. The cargo area was divided into three compartments, one for apprehended males and another for female miscreants. The two prisoner compartments were separated by a third compartment that was equipped with a sink.

Yet another spin-off of the versatile Low-Step Urban bodies came in the form of the Urban-Delivery body, offered in the customer's choice of two different models. The Urban-Delivery consisted of a Low-Step Urban cab and front clip mated to a Montpelier-produced box van body. Route drivers stepped through a large opening at the back of the cab and into a cargo compartment that was fifteen feet long and offered a full eight hundred cubic feet of cargo volume.

More than virtually all of its competitors, Montpelier Manufacturing earned a long-standing reputation for producing truck bodies that were precisely tailored to the specific demands of each customer. The company's versatility was due in large measure to the "anything that rolls on wheels" attitude of Harry Schwartz and the skilled labor force that he relied upon. The company even designed a vehicle to resemble a trolley car. Referred to as the "Jolly Trolley" when first produced, vehicles of this type are used with frequently now as a means of conveying visitors on tourist routes, everywhere from Boston to Bardstown. Unfortunately, no photos of a Jolly Trolley have surfaced.

Dark Clouds Gather

FRP bodies and the new Urban models put Montpelier in a stronger competitive position, but they were expensive to develop, and returns on the investments were not immediate. Though the future looked bright, the short-term brought cash-flow problems. At a September 12, 1957 board meeting, Harry Schwartz advised the directors that "there was a possibility that the company would not have sufficient profits at the fiscal year end, September 30, 1957, to make a contribution to the Montpelier Manufacturing Company Profit-Sharing Retirement Trust Plan" under the formula contained in the trust agreement. Given the company's financial status, the board committed to a $5,000 contribution to the plan if profits weren't sufficient to make a larger contribution in accordance with the terms of the trust agreement.

Three weeks later, the board reconvened and decided to sell its lake front cottage and surrounding land to Harry Schwartz for $10,000. No prior mention of a company-owned cottage had been noted in board meeting minutes and, obviously, no indication as to the value or usefulness of the property as a company asset had ever been put forth.

Montpelier Manufacturing described this truck their Urban Van in a 1958 sales brochure. Employee recollections of this model are extremely vague, indicative of the fact that only a few were built. There is a strong resemblance to the semi-trailers that were converted to large vans by Fageol in the early 1950's. At least one Urban Van was placed in service by the Marshall Field Department Stores in Chicago.

Sources other than Montpelier's board minutes reveal that the company purchased the cottage, located on Clear Lake in the extreme northeast corner of Indiana about twenty-five miles west of Montpelier, in 1943. When not serving as the Schwartz family retreat, the cottage was used as an occasional site for entertaining Montpelier Manufacturing customers. The directors referred to the property as a "cottage" in an apparent attempt to project an image of corporate responsibility and fiscal restraint but the dwelling and its surrounding property was far beyond the common conception of a cottage. Built into a hillside, the home appeared to be very modest when viewed from the road, the predominant feature being a three-car garage. Three bedrooms were located behind and one level below the garage. Two levels below the garage, a ground level floor faced the lake, featuring a kitchen and a great room with vast areas of glass providing a panoramic view of Clear Lake. Ample storage and additional sleeping quarters were available on the homes' fourth and highest level.

The cash payment was made in sufficient time to be entered into Montpelier Manufacturing's books of account by the end of the 1957 fiscal year. Divestiture of the cottage no doubt placed the company in a better position for acquiring bank loans or lines of credit that might be needed to sustain or expand the business. Access to credit was essential, as Schwartz had committed the company to a project that would stretch the company's financial and engineering capacities to the limit.

Just when you thought that you'd seen it all, along comes this truck. Harry Schwartz relished unusual product engineering challenges, as long as profits from Montpelier's more mundane delivery trucks allowed it. Note the grab handle on the right front "A" pillar, the diamond tread plate welded to the right side of the front bumper and the step plate just below the headlights, all of which combine to facilitate ingress and egress. The "clamshell" extending across the roof appears to connect two sliding side doors that open and close in unison.

12

Montpelier No. 75 Urban, Painting Yourself Into a Corner

In the mid-1950's, Bill Kunsman was selected to be an engineering apprentice assisting in the development of a radically different delivery truck that would break new ground by using a chassis specifically designed and built by Montpelier Manufacturing. Prior to the emergence of this new concept, all Montpelier bodied trucks utilized assorted chassis produced by the major light and medium duty truck builders.

A prototype 1958 Montpelier Mono-Lite Model 75 compact delivery sporting the same paint treatment utilized ten years earlier upon the introduction of the new line of 1948 Urban forward control delivery trucks.

The new trucks, designed under the direction of Harry Schwartz with the engineering support of Kenny Harmon, Chuck Brigle and Oscar Henry, were known as the Montpelier No. 75 Urban. The bodies were of unitized construction, similar in concept to automobiles produced by Nash in the early 1940's and Hudson in the post-WWII era. Under the skin of the outer body was a skeletal understructure welded to stiff rails beneath it. A subframe, engine, clutch, transmission, radiator, parking brake, front leaf springs, steering linkage, front axle assembly (complete with front brakes, wheels and tires) and all of the engine's electrical and belt-driven accessories were assembled into a separate module elsewhere in the plant and mated to a completed No. 75 Urban body during one of the final phases of truck assembly.

To obtain correct alignment between the engine-subframe module and the unitized truck body, tapered dowel pins on each side of the subframe were aligned with mating holes in the truck undercarriage. As the body was lowered over the subframe module, the pins engaged their mating holes and the subframe module was securely bolted to the undercarriage. If major repairs were required to the engine or transmission, the entire module could be removed in 45 minutes. Installing a replacement module in roughly the same amount of time would get the truck back on the road quickly, the driver making his routine deliveries while service work was performed on the original module.

The Model 75 Urban was developed in response to demands for improved fuel economy and as a competitive response to smaller delivery vans, primarily those that were being imported by Volkswagen. The earliest Model 75 prototypes were powered by a 75 hp 4-cylinder Ford tractor engine. A finished Model 75 with its fiberglass reinforced plastic body tipped the scales at 3,500 pounds. Anticipated payload was in the neighborhood of 2,500 pounds. The translucent nature of the unpainted FRP roof panels was again used to good advantage, providing natural illumination of the cargo area. The floor height at both of the side entry doors of the driver's compartment was at curb-level. The loading height at the rear doors was a low 24 inches, as much as twelve inches lower than some competitive trucks. Overall length of the truck was a mere 184 inches. The cargo compartment was 66 inches high, 68 inches wide and a full 96" long, providing 250 cubic feet of available cargo space.

In the April 28, 1958 edition of Montpelier's *Leader-Enterprise*, Harry Schwartz announced, "A prototype is under construction now, and should be completed before the end of the month. Plans for the new unit have been underway for nearly two years." Schwartz went on to state that, "The large truck manufacturers have failed to recognize economy is still basic to good business, and while all business costs have been climbing, trucks have become larger and more costly to operate. Frequent-stop-delivery trucks often average only 10 miles per gallon, while this new truck should operate at around 15 or 16 miles per gallon of gasoline.

"The trend to smaller and more economical delivery vehicles has become more and more apparent, and this truck should incorporate the economy of the foreign truck with the size and capacity of the American truck." Schwartz dispelled rumors that the truck would be powered by a Volkswagen engine, asserting that all driveline components, including the engine, would be American made and, thus, time proven and easily serviced.

Spurred by customer demand, Montpelier released another innovation at the time of its first public announcement of the Model No. 75 truck. Known as Dust-Control models, certain Montpelier bodies were now capable of preventing dust intrusion into the cargo area, an attribute of particular importance for bakery, laundry and milk trucks or any other delivery truck application where it was essential to maintain top quality appearance of the cargo during transit. Because delivery trucks create an area of low atmospheric pressure in their wake, dust laden air is drawn in through even the smallest gaps around doors and in between body panels along the front and sides of the truck, allowing dust to settle in the cargo compartment before air exits to the area of lower air pressure through gaps at the rear of the truck. A new venting system developed by Montpelier Manufacturing generated a positive pressure in the cargo compartment of a Dust-Control model, preventing incoming drafts and undesirable dust intrusion.

A marketing decision with positive impact was made to change the name of the Model No. 75 truck to the "Mono-Lite Model 75", a designation intended to impart awareness of the truck's "Mono" or unitized body-and-frame construction and its light weight.

Another early Mono-Lite 75 prototype with a more conventional paint treatment, unlike the three-toned Mono-Lite on the first page of this chapter.

By October 1958, six prototype Model No. 75 Urban trucks were in daily service, four of them powered by 70 horsepower Hercules air-cooled engines and the remaining two by 75 horsepower liquid cooled 4-cylinder Ford engines. Lacking a radiator and liquid engine coolant, the Hercules engine provided an extra measure of weight reduction that was one of the Mono-Lite's primary selling features. Realizing, however, that a full FRP bodied version of the Mono-Lite wouldn't have universal acceptance, Montpelier developed a steel-bodied version of the Mono-Lite on the heels of FRP development program. The company revisited the Mono-Lite's payload capacity after the first several units were built and determined that a 1,500 pound payload was the limit on a day-in day-out basis, a figure that was still more than adequate for the truck's target markets.

Starting with the preceding photograph, Studebaker skillfully grafted a 1959 Lark front end panel below the Mono-Lite's cowl. STUDEBAKER is emblazoned above the grille in individual letters. It's uncertian how far this exercise proceeded but it's almost certian that the effort was confined to the photography studio. A more conservative phtotgraphic styling excercise by Studebaker involved grafting a only Lark grille in place of the Mono-Lite's standard grille, leaving the rest of the Mono-Lite front clip intact. A photo of that truck has been published elsewhere but was not available for this book.

The easily removable engine and front suspension module was a feature that prospective buyers of Mono-Lite trucks looked forward to with great anticipation. One of Montpelier's existing fleet customers indicated that this feature alone would enable him to cut in half the number of spare trucks that he kept on standby to fill in for trucks undergoing repairs. A very desirable and worthwhile feature, the removable power module wasn't a new innovation but instead a reprise of a feature that had been offered years earlier by Pak-Age-Car, White Horse and Marmon-Herrington Delivr-All forward control delivery trucks.

As field testing of the six trucks steadily established the Mono-Lite's reputation as an economical and very functional truck, Schwartz emphasized to a growing list of prospective customers that "We came to this design by first considering the driver and his comfort and convenience and then the accessibility and workability of the load. Next we positioned the chassis components. This meant we had to ignore the old method of using the truck manufacturer's standard chassis and building a delivery truck around that." Engine selection for production Mono-Lites was still an open issue, even as the six prototypes were in daily field test operation.

A rear view of the Mono-Lite Model 75 reveals a smooth, clean, parallel sided interior, a step bumer and rear doors that provide full width access to the cargo compartment and swing out against the sides of the truck to facilitate cargo transfer at a loading dock. Plenty of natural light passes through the unpainted portion of the FRP roof but a cargo compartment light is available for night time operation. Unlike Montpelier's traditional delivery truck bodies, Mono-Lite Model 75 trucks did not use a chassis produced by one of the major light-duty truck manfacturers. Mono-Lites were of unitized contruction, built at the Montpelier plant form the ground up.

U. S. market demand for forward control delivery trucks in the under 10,000 pound GVW class in 1957 amounted to over 21,000 trucks, a fact that prompted Montpelier's traditional rival, International, to introduce a small delivery truck that would be a direct competitor to Montpelier's Mono-Lite. International's Metro-Mite AM-80, introduced in the Fall of 1958, was positioned below Montpelier's Mono-Lite 75 in terms of overall size, cargo volume, payload and horsepower.

It bears repeating that Mono-Lite bodies were of unitized construction consisting of a skeletal understructure welded to stiff steel support rails. The simplicity that came with years of ordering a pre-engineered forward control truck chassis consisting of a thousand or more parts as a single purchased item, economically manufactured and assembled by any one of several high volume truck manufacturers, was gone. Instead, Montpelier's Purchasing Department was confronted with the task of purchasing every mechanical component of a Mono-Lite 75 truck, from cotter pins and clutches to radiators and rear axles. The economies of large scale production simply didn't exist at Montpelier Manufacturing and Bob Treer, Montpelier Manufacturing's experienced and capable purchasing agent, was pressed beyond measure in his efforts to obtain the many piecemeal parts and components that comprised a Mono-Lite 75 truck at anything approaching reasonable cost. Montpelier's traditional chassis suppliers viewed the Mono-Lite 75 as a competitive assault on chassis sales and greeted Montpelier's requests for pricing on various driveline and chassis components rather indifferently. Of the price quotations that Montpelier received, most were disappointingly high. Shunned for the most part by high volume low cost component suppliers, Montpelier resorted to smaller companies or companies that weren't directly or indirectly competing with Montpelier in the light duty delivery truck market.

Harry Schwartz was caught totally off his guard by the cold reception Montpelier Manufacturing received from the high volume producers of light duty truck chassis and driveline components when price quotations were solicited. Over the many years of his association with the Big Three, in particular General Motors, he had developed close ties to many high ranking engineering executives. Harry's son Jim recalls more than one occasion when his father was brought in by various General Motors engineering groups as a consultant to analyze and solve engineering problems that had stymied resident GM engineers. As the major players in the automotive industry spurned Montpelier's requests for quotes, Harry became incensed by the fact that he wasn't reaping the slightest dividend in return for all of the political capital, favors and good will he had invested during his many years of close collaboration with his counterparts at the Big Three. Years later, Jim Schwartz recalls his father's strong convictions that General Motors had double-crossed him at a time when he was irrevocably committed to putting the Mono-Lite 75 into production.

The October 6, 1958 edition of Automotive News, an automotive industry trade journal, covered both the Metro-Mite AM-80 and the Mono-Lite 75 in a single article, pointing out that both trucks were designed to deliver goods at operating costs that would undercut emerging foreign competitors. Driveline

The cover page of International's Metro-Mite 80 Compact forward control delivery truck for 1959. Note how closely the paint treatment matched that of the special 1948 Montpelier Urban and the 1958 Montpelier Mono-Lite 75 prototype. The similarity signals an apparent attempt by International to divert attention away from the Montpelier's Mono-Lite 75 and toward the Metro-Mite 80.

International introduced its Metro-Mite 80 compact delivery truck for the 1959 model year. Like Montpelier's Mono-Lite 75, the Metro-Mite 80 had a unitized body and frame. Offered at a base price of $2251, Metro-Mite 80s were initially powered by a high revving 51 horsepower 91 CID imported four-cylinder engine. Much needed higher output powerplants would be offered as extra cost options a few years later.

and chassis specifications for International's Metro-Mite were fairly concise and straightforward. All that Automotive News could report about the Montpelier Mono-Lite 75 was that "The engine compartment will accommodate any of a number of engine and transmission assembles. The same applies to axles." Base price for International's Metro-Mite at the time the *Automotive News* article was published was $2,050. No comparable base price was indicated for the Mono-Lite 75. Fifteen months later, the January 1960 published base price of a Mono-Lite 75 was $3059. The fifteen-month span that followed the Automotive News article was a turning point in Montpelier Manufacturing history

One factor that cast a dark cloud over Montpelier Manufacturing in the late 1950's was the aforementioned difficulty that the company had in acquiring the many individual parts and components that comprised a Mono-Lite 75 at reasonable cost. Harry Schwartz was confident enough in his innovative engineering talent to proceed with design, prototyping and production tooling for the Mono-Lite 75 truck without conducting a verifiable cost analysis at one or more intervals as the program progressed. Referring to a dedicated area of the manufacturing floor where Harry experimented with future delivery body designs, Harry's son Jim recalls his father saying that "As long as I have five trucks going out the front door, I can afford to play back there." Harry's engineering exploits brought him far more personal satisfaction than any amount of wealth he might have acquired from his business endeavors.

On top of the high component costs of the Mono-Lite 75 were the dollars invested in engineering a vehicle from the ground up, building and testing of prototypes, and tooling up for what was hoped would be a significant portion of an emerging market for small delivery trucks that were more economical to operate on a daily basis. All of the company's profits from its traditional line of Urban and Side-Aisle trucks were invested in the Mono-Lite program, not to mention funds obtained through a line of credit from two local banks in September 1958.

A third factor beyond the control of the company was the nationwide recession of 1958, a year that was especially hard on all facets of the automotive industry. Depressed sales of delivery trucks equipped with Montpelier's mature and profitable lineup of Urban and Side-Aisle delivery bodies didn't fortify the company's balance sheet as in the recent past. The emerging market for compact delivery trucks that totaled 21,000 sales in 1957 fell precipitously in 1958, a percentage drop exceeding that of all other segments of the vocational truck market.

Added to the unremitting strain brought on by these factors were three sudden and pivotal events in the lives of the Schwartz family as well as at Montpelier Manufacturing that occurred over a seven week period in late summer and early fall of 1959. Only time would tell if the portent of these events would solidify the company's future or if the company's underpinnings, so painstakingly built over more than three decades, would begin to fatigue.

Several delivery truck body manufacturers introduced compact forward control delivery trucks designs from the late 1950s onward. Their hopes for a promising niche market were dashed when Ford introduced its Econoline compact vans in 1961. With a base price of $1956, Econolines were far more affordable than Montpelier's Mono-Lite 75, International's Metro-Mite and similar trucks offered by a host of truck body manufacturers. Econolines were based on the Ford Falcon platform, enabling Ford to produce them very economically. Econoline buyers were very pleased with the truck's features and its low operating and repair costs.

13
Burdens Moved to Broader Shoulders

On Wednesday, September 9, 1959, Mrs. Grace Hill, widowed wife of J. D Hill and mother-in-law of Harry Schwartz, passed away at her prestigious 201 East Main Street residence in Montpelier after a long illness. Mrs. Hill was Chairman of the Board and a vice-president of Montpelier Manufacturing, succeeding her husband in these capacities from his death in 1939 until illness forced her to step aside.

Montpelier was still recovering from the shock of Mrs. Hill's death when, three weeks later, Harry Schwartz announced that Montpelier Manufacturing had been sold to White Motor Company. The deal was consummated after six months of negotiations and was prompted by White's interest in Montpelier's new and innovative Mono-Lite 75 delivery truck, which, for some reason, had been renamed again, this time it was called the "Series 75."

A joint development program bewteen Montpelier Manufacturing and Gould Battery produced six electrically powered Urban Delivery trucks in the late 1950's. A Gould nameplate is mounted above the grille which is closed off to air flow normally required for engine cooling. The word "Electric" is embossed into the truck's hubcaps. The hinged panel to the right of the side door allows for battery access.

Upon finalizing the purchase, J. N. Bauman, White's president, emphasized that the Series 75 was far superior in its design relative to competitive economy trucks. Montpelier had trounced its competitors by offering a removable engine module, a low driver's compartment floor and fiberglass reinforced plastic body construction. Sales of the Series 75 would be channeled through White's 200 dealers as a White product, opening a new market for White dealers that ideally complemented White's existing line of trucks and semi-tractors.

Montpelier's October 1, 1959 edition of the *Leader-Enterprise* enthusiastically reported that, "Mr. Bauman pointed out that his company anticipates a large scale of the Series 75, and this may require an expansion of the local plant after mass production gets underway, probably sometime next year." Bauman gave assurances that all existing Montpelier employees would continue in their existing functions and that Harry Schwartz would continue to manage what would now be known as the Montpelier Division of White Motor Company. Montpelier's traditional line of delivery trucks, mounted on chassis supplied by a variety of U.S. truck manufacturers, would continue to be marketed by the Montpelier Division's existing in-house sales department, i.e., totally independent of the White dealer network.

The front page news article provided further assurances of good times ahead for the Montpelier Division by reporting that "Mr. Schwartz told the *Leader-Enterprise* he expects the acquisition of the Montpelier Manufacturing Company by White Motor to be a great thing for the company's employees and for Montpelier. The company will stay here, he said, and will continue to grow. Mr. Bauman agreed and said his company hopes addition of the Series 75 line will help expand sales and enhance White's competitive position."

In thirty-two years of operation, Montpelier Manufacturing's annual sales had yet to reach the $3 million dollar level. In contrast, White Motor Company's annual sales exceeded $325 million, an indication that the Montpelier acquisition, although highly desirable and enabling White to fill a strategic and desirable market niche, was not one that would add significantly to White's growth and earnings potential.

At the time of the sale, Montpelier Manufacturing was a closely held corporation, owned by Harry Schwartz, his wife Margaret and the estate of the late Mrs. J. D. Hill. Just as Ford and GM had squeezed the independent auto manufacturers with a hotly contested automobile sales blitz in 1954, Harry felt that similar pressures (vertically integrated delivery truck manufacture, stubborn resistance to selling him components at a reasonable price) were coming his way from the same direction. Harry perceived that "the handwriting was on the wall" and it was time to sell. The sales price was not disclosed.

At the time Montpelier Manufacturing introduced the Series 75 delivery truck, there was no other delivery truck in the industry that possessed all of its outstanding features. A choice of steel or fiberglass reinforced plastic body construction, an easily removable power and front suspension module and 1500 pound payload capacity in a compact and fuel efficient package would seemingly insure that the truck would be in high demand from established as well as new Montpelier customers. Stymied by the complexities of building a truck from scratch, the indifference of the predominant driveline and chassis component manufacturers in supplying Montpelier with their products, and insufficient financial resources to tool up for volume production, White Motor Company's acquisition of Montpelier Manufacturing appeared to be the "magic pill" that would cure all of Montpelier Manufacturing's ailments. Surely the resources of a decades old company more than one hundred times the size of Montpelier Manufacturing would take whatever measures necessary to finish the job of bringing the Series 75 truck to full production while insuring that Montpelier's profitable lineup of Low-Step Urbans and Side-Aisle delivery trucks were available to the many customers that insisted upon them.

The third event that shook the foundations of the newly created Montpelier Division as well as the Schwartz family was the unexpected passing of Margaret Schwartz at age 60 on Tuesday, October 27, a mere seven weeks following the death of her mother, Mrs. J. D. Hill. Margaret Hill Schwartz died of an apparent heart attack while returning to Montpelier with her husband Harry, shortly after leaving the office of a Toledo physician, one day after the couple's thirtieth wedding anniversary.

A determined Harry Schwartz pressed onward in his traditional firm command as president of the Montpelier Division of White Motor Company. An expanded line of PDQ trucks was under development (yes, the Model 75 Urban/Mono-Lite 75/Series 75 trucks had been renamed once again - the list of designations now numbered four). Larger versions of the PDQ (**P**ickup-**D**eliver-**Q**uick in

Instrumentation and driver controls in the Gould electric trucks are few. Wheelhousings were extended to enclose a series of batteries supported by plates and outriggers welded to the frame.

White Motor Company's lexicon) were developed precisely as Bauman had indicated at the time of White's takeover of Montpelier Manufacturing. The Montpelier Division was tailoring the trucks to the specific needs of individual customers, just as they had done for many years with their traditional Urban, Super Urban and Side-Aisle delivery bodies. In May 1961, six PDQ units were configured as ambulances and sold to the Cleveland police department. Following completion of that order, the Montpelier Division produced 50 White PDQs for Nickles Bakery. H. J. Nave, President of the White Truck Division, announced that White PDQs, as well as a suitable range of trucks with Montpelier's established line of delivery bodies, would be featured in the Baking Industry Exposition to be held in Atlantic City in October.

White's purchase of the company and the increased influence it wielded with component suppliers relieved at least some of Montpelier's earlier problems in getting component manufacturers to agree to supply the Montpelier Division with PDQ components. PDQs for the 1960 model year could be had with the customer's choice of a rugged 134 cubic inch displacement four-cylinder Willys F-head Hurricane engine or a six-cylinder Continental overhead valve engine of 226 cubic inch displacement. A generously sized 10½-inch diameter clutch transmitted torque through a three-speed full synchromesh transmission to a 5.87:1 rear axle. Engine options grew to three in the following year with the addition of a four-cylinder overhead valve Continental engine of 193 cubic inch displacement.

Securing supplier commitments was a positive development but it only served to magnify the next problem that the Montpelier Division faced in bringing the PDQ line of trucks into full production. The logistics challenges that Bob Treer, Montpelier's purchasing agent, faced in ordering and maintaining inventories of the many individual parts and components that comprised a PDQ truck have already been pointed out. With all of the parts and components flowing into the plant, the daunting task of manufacturing unfamiliar unitized bodies and equally unfamiliar front suspension and power modules fell on the shop superintendents, foremen, and shop floor employees. Components such as front and rear axles, transmissions, engines, clutches, radiators, springs, steering boxes, all had to be properly assembled and perform just as efficiently and reliably as the time-proven

Battery enclosures and an elevated cargo floor reduced the available cargo volume of the Gould battery-powered Urban trucks. The absence of a driveshaft tunnel resulted in a totally flat floor in the driver's compartment. A partition with sliding door separated the cargo and driver's compartments on the truck in the lower photo.

chassis assembled by more experienced manufacturers for Montpelier's traditional line of trucks. Further, the company had to continue production of its traditional Urban and Side-Aisle trucks, each tailored to specific customer requirements, while building PDQ trucks *without any increase in plant equipment or floor space.*

PDQ supplier recruitment, logistics, and manufacturing problems would have been manageable for a hardworking core of shop and salaried employees if the Montpelier Division could have kept its focus on the development of one compact White PDQ model and its associated options before expanding upon the PDQ concept. But immediate gratification overruled common sense at White Motor Company's Cleveland headquarters and the Montpelier Division was goaded into quickly expanding the PDQ lineup into several series. Within a few years of the launch of the initial compact truck conceived by Harry Schwartz, White PDQ's were available with cargo compartments as long as fifteen feet and a bewildering assortment of body heights and widths. The Cleveland contingent bore down hard on its Montpelier appendage and the resources that were expended on the White PDQ product line (which had yet to turn a profit) came at the expense of the profitable but forcibly neglected line of Montpelier Urbans and Side-Aisle delivery trucks.

As they had done for years, Montpelier continued to produce bodies that were installed on Chevrolet, GMC, Ford, Dodge or, occasionally, International and Studebaker chassis. These trucks logically bore the nameplates and other identifying marks of the chassis manufacturer, not White Motor Company. White was irked by the popularity of these trucks and the fact that they competed directly against the White PDQ. The fact that Montpelier Urbans and Side-Aisles were assembled in a plant that was a division of the White Motor Company made the arrangement all the more exasperating. Adding to White's frustration and disdain was the fact that Montpelier bodies installed on competitors' chassis were *profitable* while the built-from-scratch White PDQ's were not.

Although White kept the Montpelier Division on a tight leash, Harry did find time to collaborate with the Gould Battery Company on the production of six small battery-powered Urban delivery trucks. Montpelier employee Denver Miller recalls that Gould provided technical support and that Harry Schwartz engineered the vehicles, in particular the motor layout and the structural modifications required to support the weight of the batteries. Montpelier kept the trucks for roughly six months, driving them over progressively longer stop-and-go routes to get a feel for the range and overall performance of the trucks. They were then delivered to *The Philadelphia Inquirer* for what was hoped to be routine newspaper distribution. *The Inquirer* found the trucks to be significantly underpowered in relation to their weight and incapable of negotiating hills at a reasonable rate of speed. It isn't known how long the trucks remained in service or what ultimately became of them but the joint project with Gould proceeded no further.

White's other Interests

The PDQ was by no means White Motor Company's first venture into light and medium duty delivery truck production. White Motor Company offered panel trucks in the 1920's and released their own version of Montpelier's Side-Aisle delivery truck in 1930. Over time, however, the company seemed to repeatedly lose its grip on the light and medium duty truck markets and retrenched into the production of heavy-duty trucks where it seemed to have much more staying power among its competitors. White's introduction of a new line of forward control delivery trucks in 1939 competed directly with Montpelier's Urbans and the growing number of other manufacturers producing similar trucks. Marketed as the White Horse, sales of the new line took off like a quarter horse with a swarm of hornets in hot pursuit. Dairies, department stores and laundries in major metropolitan areas bought scores of the air-cooled, rear-engined White Horse trucks, some fleets operating as many as two hundred units. By 1947 fleets realized that the quirky rear-engine White Horse had little to offer in relation to more conventional forward control delivery trucks and the pre-war White Horse was consigned to oblivion. Replacing it was a new White Horse model that was an overcorrection to the foibles of the pre-war model. Far from being an efficiently packaged forward control truck, its "dog house" engine compartment was well outside the envelope of the delivery truck body, a protrusion that made Divco's much smaller proboscis look decidedly pug-nosed. The post-war White Horse trotted off into the sunset, just as its predecessor did.

To say the least, White Motor Company was a dominant competitor in the agricultural and commercial truck markets. An established medium and heavy duty truck manufacturer in its own right, White went on a buying spree in the eight years preceding the purchase of Montpelier Manufacturing, buying out Sterling, AutoCar, Reo, and Diamond-T, four of it's long-time heavy-duty truck competitors. Acquisitions didn't stop with the Montpelier Manufacturing purchase. White also began buying up agricultural equipment manufacturers in the early 1960's. On the heels of their acquisition of the Oliver Farm Equipment Company and Cockshutt

International Metro-Mite AM-80 vs. White PDQ
Comparative Specifications for 1960 Models

	International Metro-Mite AM-80	White PDQ A8
Wheelbase (in.)	96	104
Overall Length (in.)	159	188
Standard Engine Configuration	Internatinal A-55, In-line 4-cylinder	Willys 4-cylinder F-head
Horsepower	51 @ 4250 RPM	70 @ 4000 RPM
Body Construction	Unitized, All Steel	Unitized, FRP and Steel
Cargo Dimensions		
Height (in.)	66	62½
Width (in.)	66	63 ½
Length (in.)	84	93⅛
Cargo Volume (cu. ft.)	200	207
Rear Door Lift Height (in.)	25	24
Rear Door Height	57½	56
Rear Door Width	38	66
Truck Weight	2800	4150
Payload Capacity (lb.)	1000	850-2600
Base Price	$2,251	$3,059

Farm Equipment of Canada in 1960, White added Minneapolis-Moline to its agricultural equipment portfolio in 1963. White also had lucrative agreements to market and distribute heavy-duty long haul trucks produced by Freightliner and Western Star through its far-flung network of White dealers. Seeking a captive source for heavy-duty diesel engines, a lengthy courtship with Cummins Engine Company in the early 1960's failed to result in a merger. Each company remained separate, continuing a long-standing customer-supplier relationship that went back nearly three decades.

White's post-war history can be likened to the Midas touch in reverse. Their acquisitions and marketing agreements should have given them a commanding share of the truck and agricultural equipment markets and perhaps they did for brief periods after the acquisitions were completed. However, a review of White's share of the U.S. light, medium and heavy-duty truck market over a fifteen year span from 1947 to 1962 reveals something altogether different. The table on these pages reveals a precipitous drop in White's market penetration in spite of their growth-by-acquisition strategy.

In 1947, White enjoyed a commanding share of the market for trucks in the 16,000 lb. GVW and higher classifications. By 1962, their presence in each of these classes had significantly deteriorated or vanished altogether. The 400 small trucks carrying the White banner in 1962 were White PDQ's that rolled out of the Montpelier Division. If White's competitors were keeping track of these figures, it's certain that they tipped their hats to White at the close of each model or fiscal year, thanking White for faithfully handing them a few more percentage points of market share.

Industry professionals that had occasion to work with White Motor Company in the 1960's and later years watched in amazement as White bought smaller companies that were on the verge of bankruptcy and simply closed the businesses down. The owners of these smaller businesses, friends and associates of White's top executives, were thrilled by the prospect of avoiding bankruptcy and leaving their foundering businesses with generous monetary consolation prizes, courtesy of White Motor Company.

White's factories were antiquated in relation to their competitors. For the most part, Montpelier Manufacturing's physical plant was antiquated too, a fact that Montpelier's shop employees overcame with pride, hard work and a good deal of blood and sweat in their efforts to stay competitive. If the Montpelier facility had been state-of-the-art for 1959, perhaps White might have determined that Montpelier Manufacturing didn't fit into the White mold and White may have pursued another opportunity. If, on the other hand, White were given the opportunity to transform the Montpelier Division into a state-of-the-art facility would they do it? In due time, the opportunity, and White's response, would be forthcoming.

This table illustrates the drastic decline in market share that White Motor Company incurred over a fifteen year period. Even though White bought out four competitors during this period, it's temporary increases in market share were fleeting and unable to reverse a steady and precipitous decline. White Motor company had an engineering corps that was very adept and highly respected throughout the truck manufacturing industry but antiquated plants and a stumblebumb corps of top excutives would have been an interesting case study for Scott Adams, syndicated cartoonist for the "Dilbert" comic strip.

Production Volume and Percent Market Penetration in Various Commercial Truck Classifications for White Motor Company and Affiliates[1], 1947 and 1962

	1947		1962	
Truck Classification	Units	Percent	Units	Percent
Under 16,000 Lb. GVW	18,384	2.28	400	0.05
16,001 to 19,500 Lb. GVW	5,379	21.31	67	0.06
19,501 to 26,000 Lb. GVW	10,035	34.51	1,417	2.04
26001 to 33,000 Lb. GVW	7,584	38.93	15,087	22.47

[1] *Includes Autocar, Diamond T, Sterling and REO*

14
New Management, New Challenges

With the production of the White PDQ gaining momentum and the transfer of ownership and oversight of Montpelier Manufacturing successfully completed, Harry Schwartz felt assured that prosperity lay ahead for the Montpelier Division. He had achieved virtually everything he strived for in his 34 years at Montpelier, and in June of 1961 Schwartz announced his retirement.

In the two years that followed White's buyout, Harry had felt constrained by White's hierarchy but he accepted it as an inevitable consequence of the sale. Privately, he told his family and a few close friends, "When you've been the president and the driving force for this thing for many years and all of a sudden you are purchased by someone else, the first thing they do is send someone in and take over the finances which is very natural."

After White Motor Company bought Montpelier Manufacturing, it freshened up Montpelier's triple-toned Mono-Lite 75 display truck and added "White PDQ" badging to kick off a marketing effort for its newly acquired line of compact forward control delivery trucks.

In ceding financial control of the company to White, Harry said, "That's fine. I'll run the company for them." White wanted Harry to run the Montpelier Division for them because he knew the operation like no one else and was uniquely capable of guiding the company in a direction that suited White's interests. White tacitly entertained thoughts of moving PDQ production elsewhere but kept the operation in Montpelier because it was too difficult and costly to tool up for PDQ production elsewhere. Harry was gratified that he still maintained control over engineering, his first love and lifelong interest. Eventually, White wanted to increase its involvement in the engineering of new and existing products and, when that became apparent, Harry concluded, "If they take that away from me, then they better do the whole thing themselves." At age 65, he was ready to leave and did so without any bitterness or ill feelings, fully understanding that the company that was formerly his should rightfully be managed by its new owners.

As reported in the June 5, 1961 edition of Montpelier's weekly newspaper, "Mr. Schwartz told the *Leader-Enterprise* that his association with White has been 'very amicable and friendly.' He said he is confident the local division will 'really go places' under the new management." After years of intense effort and devotion to charting Montpelier Manufacturing's success, Schwartz was certainly entitled to a well-deserved retirement. However, as the inveterate, innovative, and industrious founder, owner and leader of Montpelier Manufacturing since its inception, it's hard to imagine him submitting to White's management hierarchy in docile corporate fashion. If all were spinning in a positive direction upon his retirement, the statement that his association with White had been, "Very amicable and friendly," would seem totally unnecessary. Instead, the statement served as an invitation to read between the lines of his retirement announcement. If he had been in absolute control in the two years following White's takeover, it's doubtful that he would have allowed the pendulum to swing so far in the direction of White's PDQs and away from Montpelier's profitable Urban and Side-Aisle trucks.

Pictured beside the C Series White PDQ truck is a PDQ engine and front suspension module. This module is equipped with a 190 CID Continental OHV four-cylinder engine that was offered in White PDQ's for two years, starting in 1961. Delivery fleets were encouraged to dispose of older delivery trucks that were kept as spares and instead keep a spare PDQ module on hand if major engine repairs were necessary. White assured customers that a module replacement took no more than three hours to accomplish.

This photo was one of several in a White Motor Company salesman's kit that was used to acquaint customers with the many desirable features of White PDQ forward control delivery trucks.

The immediate impact of Schwartz's retirement on the future of the Montpelier Division was open to much speculation, but the division's employees viewed his departure in an ominous light. Even though Schwartz had clearly not been pulling all of the strings over the preceding two years, many employees still looked upon him as the ever-present commander-in-chief. The anguish experienced by some of the rank and file employees after one of Schwartz's abusive verbal tirades was tempered by the fact that they knew he was always out in front of them defending the business against threats to its success and to their continued employment.

Roy J. Sjoberg, who had joined the Montpelier Division only a few months earlier as general sales manager, was promptly chosen as Schwartz's successor and General Manager of the Montpelier Division. Before joining White, Sjoberg had spent nine years as vice-president of sales at Divco, one of Montpelier Manufacturing's long-time competitors.

White also turned to the Montpelier Division's sales force to select an operations manager. John C. Rundell, the Montpelier citizen who organized the lavish "E" Award ceremony for Montpelier Manufacturing in August of 1944, had joined the company in 1947. After a fourteen year stint in Montpelier's labor relations and sales departments, Rundell was serving as Montpelier's Vice President in Charge of Sales when he was picked to be the division's new operations manager. Dividing Harry Schwartz's responsibilities between two successors was an admission, both by White Motor Company and the Montpelier Division, that it would take more than one man to fill Harry's shoes.

White Motor Company retained Schwartz as an independent consultant for some time after his retirement. Within weeks of his retirement announcement, he left for the west coast, reportedly in search of additional manufacturing facilities for PDQ trucks. Apart from his sometimes incendiary approach to running Montpelier Manufacturing, Schwartz was a highly respected and admired man in the community, possessing what many described as a heart of gold. He had served on the Montpelier School Board, was a charter member of the local Rotary, active in the Masonic Lodge and devoted much time and effort to his special project, the building of the Williams County General Hospital. All of Montpelier recognized and appreciated his community spirit and would sorely miss his sincere concern for the well-being of all Montpelierites.

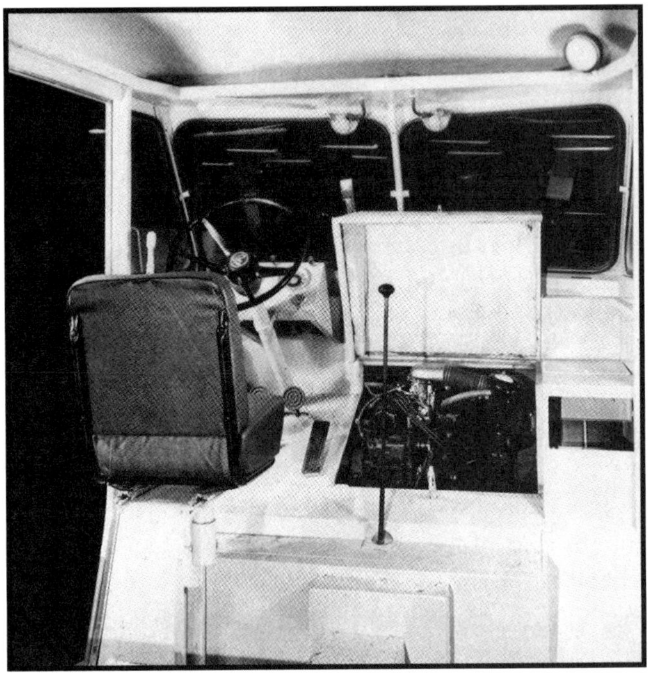

The driver's working area for a White PDQ. Brake, clutch and accelerator pedals appear to be high in relation to the driver's seat, forcing the driver to extend his legs into an uncomfortable horizontal position while driving. Access to the four-cylinder Willys Hurricane engine is very good. The absence of framing in the roof of this PDQ indicates that it was built with a steel top rather than the highly touted FRP alternative.

Schwartz's retirement was by no means the only event that undermined employee confidence in the future of the business and their job security. The positive spin coming from statements made by Harry Schwartz and White's J.N. Bauman when White Motor Company assumed ownership lost their credibility in the months that followed the sale. Bauman's October 1959 reference to "... an expansion of the local plant after mass production gets underway, probably sometime next year," never materialized. In truth, the plant was in need of a total modernization, not merely expansion. By the 1960s, the antiquated wooden floors were either oil soaked or covered with layers of paint, undercoating, or plastic resin. There was hardly a level spot in the plant, a fact that was frustrating to shop floor workers in their efforts to "eyeball" a truck body to see if everything was plumb and square. Occasionally the aging planks would give way to the weight of a truck body and the cart on which it was mounted. Workers would set their tools aside, pry the body and cart out of its freshly-made crater and make spot repairs to the floor. Employees kept a watchful eye on the floors wherever welding or cutting torches were in use, doing their best to extinguish the occasional smoldering floorboard. The crawl space under the factory floor was a veritable wind tunnel, capable of fanning a glowing ember into a blazing fire, if given the chance.

The incompatible corporate cultures and the fact that White Motor Company didn't reach into its presumably deep pockets to invest in the business as J. N. Bauman had hinted left some Montpelier employees feeling left out in the cold. White was true to its word in assuring that all of Montpelier's existing employees could remain if they wished, but an exodus of sales, engineering and administrative employees began after White assumed ownership, primarily due to doubts about long term job security.

While Schwartz's retirement may have had negative connotations for some existing Montpelier employees, it was viewed as a positive opportunity for at least one former employee. Bill Sayers had left his sales position at Montpelier Manufacturing in 1953 to become the fleet sales manager at the DeKalb Commercial Body Corporation, a Montpelier competitor in DeKalb, Illinois. Sayers left his position at DeKalb and returned to his native Montpelier to become the Montpelier Division's assistant general manager in 1963. Bill's sales experience at one of Montpelier's prime competitors was helpful in ironing out some of the rough spots in the White-Montpelier working relationship, if only for a short while. Bill's return to Montpelier ended prematurely in May 1965 when, at the age of 41, he was stricken with a fatal heart attack in the presence of his Montpelier Division coworkers.

Roy Sjoberg's tenure at the helm of the Montpelier Division lasted little more than a year. No reasons for his July 1962 departure were offered. He left Montpelier to join the staff of H. D. Weller, White Motor Company's vice-resident of sales, at White's corporate headquarters in Cleveland, supposedly to devote his time to PDQ sales. Casual observations point to the fact that he was obviously a Montpelier outsider who had to fill the shoes of a man who had run the company for decades in a uniquely stern and forceful manner. In addition to being unfamiliar with the Montpelier work force and way of life, it is certain that his more conventional style of management was totally foreign to his subordinates. He also may have shared the attitude of other White corporate executives who sojourned from their lofty Cleveland headquarters for an occasional Montpelier visit. For them, a brief visit was fine, but a dispatch from the corporate offices to the remote reaches of Ohio for more than a few days was disdainful. The majority of Montpelier shop and office employees weren't exactly smitten with their occasional White visitors either. Arriving in fancy cars and clothed in fine suits, they shunned the shop floor and Montpelier employees felt disaffected by their hit-and-run visits.

With Sjoberg's departure, John Rundell was elevated to the position of general manager, a move that was warmly welcomed by the majority of Montpelier Division employees. Rundell anticipated a good future for the Montpelier Division and planned no changes to the division's various department heads upon assuming his new responsibilities. In a July 12, 1962 *Leader-Enterprise* article announcing the

Nine White PDQ A Series trucks destined for Nickles Bakery, part of a much larger order that Nickles placed for these trucks.

management changes, Rundell affirmed that, "I have the utmost confidence in these men, and together we will make a success of this operation."

Various department heads serving the Montpelier Division at the time included:

Sales Manager:	**Guy Priest, Jr.**
Controller:	**Robert Lester**
Director of Planning:	**Harold Bartick**
Chief Engineer:	**Glen Taylor**
Shop Superintendent:	**Henry Pape**
Purchasing Agent:	**Robert Treer**
Quality Control:	**William Sprankell**
Shipping & Receiving:	**Art Donaldson**
Parts and Service:	**Robert Poucher**
Manufacturing Department Heads:	**Denver Miller**
	Paul Strawser
	George Cornell
	Jay Ransom
	Robert Boszor
	Adam Alquire
	William Kunsman

Ferman Haase, former plant superintendent, retired in February 1962 and was serving on a consultant basis. At the time Rundell assumed responsibilities as general manager, the Montpelier Division had an order for 75 Railway Express Agency units and the overall truck order backlog was greater than at any time in the preceding year.

It wasn't long before Sjoberg either flew from or fell off his perch at White's Cleveland headquarters, retreating to Divco's Detroit headquarters and assuming the position of sales manager. Sjoberg and Divco became involved with the Field Body Company of El Monte, California on a program to install FRP bodies on Divco chassis equipped with the standard Divco windshield and front clip. The Montpelier Division was apparently overlooked as a potential supplier of Divco's FRP bodies, in spite of its closer proximity to Detroit and a more central location for nationwide distribution. Several factors could have ruled out a collaborative effort between Divco and the Montpelier Division. Divco and Montpelier had been direct competitors for over thirty years and Sjoberg may have had considerable difficulty resuming a relationship with White after returning to a White competitor. White may have had no inclination to dilute their competitive advantage by sharing FRP bodies with a competitor and the Montpelier Division couldn't risk devoting FRP production capacity to Divco at the possible expense of existing commitments for PDQ and Montpelier bodies.

In spite of the turmoil, tragedy and turnover, the PDQ line was significantly expanded in accordance with White's demands. By 1964, the White PDQ line of forward control delivery trucks was offered in a full range of body sizes. White PDQ's of all sizes could be found in a wide range of delivery truck applications, as evidenced by photos found elsewhere on these pages. John Rundell noted that, "The White PDQ covers the entire multi-stop delivery market with a series for every use and adaptations for every vocation. Load spaces come in four heights, four widths and lengths from seven to fifteen feet." PDQ bodies were also fabricated from traditional steel panels for customers who felt that a steel-bodied truck was better suited to their modes of operation. Complete specifications for PDQ Series A, Series B and Series M trucks for the 1961 model year can be found elsewhere on these pages.

Chrysler Corporation Slant Six and 318 CID V-8 engines took over as the predominant engines for White PDQ trucks for the 1963 model year as the Willys and Continental engines were older, heavier and more costly designs in steady state of decline as mainstream automotive powerplants. Saginaw Gear supplied front axles and steering boxes while tire purchases were spread among several suppliers. Warner Clutch and Brake was the preferred clutch supplier and Dana Corporation supplied PDQ driveshafts and 4.88:1 rear axles. The increased horsepower and torque of the Chrysler engines made the change from the previous 5.87:1 rear axle ratio feasible and enabled the trucks to operate at higher cruising speeds without sacrificing acceleration in stop-and-go traffic. A Perkins 4-236 four-cylinder diesel engine joined the list of available engines, presumably at White Motor Company's insistence. Customers willing to sacrifice the easy starting and soot-free exhaust of gasoline engines could pay a premium for the improved

Specifications for A, B, C and M8 Series White PDQ Trucks												
	A Series		B Series			C Series			M8 Series			
	A7	A8	B8	B10	B12	C10	C12	C15	M8-7	M8-8	M8-10S	M8-12D
Wheelbase	96	104	104	122	137	122	137	154	104			
Overall Length (in.)	179 7/8	187 7/8	187 7/8	216 7/8	240 7/8	216 7/8	240 7/8	276 7/8	191			
Overall Height (in.)	87½		93			103½		105½	97	98		
Overall Width (in.)	75¼		79 ¾			87			79¾			86¾
Gross Vehicle Weight (lb.)	5,000-7,000		6,000-7,000		7,000	7,000-10,000		10,000	7,000	8000	9000	10000
Net Weight (lb.)	4,150	4,250	4,350	4,550	4,900	5,330	5,740	6,780	5,200	5500	5700	5900
Payload (lb.)	850-2,600	750-2,500	1,650-2,450	1,450-2,300	2,100	1,670-4,500	1,260-4,200	3,220	2,000	3000	4000	4700
Cab												
Inside Height (in.)	75		79			85			81			
Step Height (in.)	13		14			15½		17	15	15¾		
Door Opening (in.)	29¼		29¼			29¼			33			
Door Height (in.)	65		69			70			71			
Standard Engine												
Cylinders	4, F-head		4, F-head		4, OHV	4, OHV		6, OHV	4, OHV	6, OHV		
Displacement (cu. in.)	134		134		193	193		226	193	226		
Horsepower	70 @ 4000 RPM		70 @ 4000 RPM		78 @ 3200 RPM	78 @ 3200 RPM		110 @ 3200 RPM	78 @ 3200 RPM	117 @ 3500 RPM		
Peak Torque (lb. ft.)	111 @ 2000 RPM		111 @ 2000 RPM		171 @ 1600 RPM	171 @ 1600 RPM		192 @ 1800 RPM	171 @ 1600 RPM	192 @ 1800 RPM		
Transmission	Heavy Duty 3-speed Synchromesh											
Clutch diameter (in.)	10½		10½		11	11			11			
Electrical System Voltage	12		12			12			12			
Generator (amp)	35		35			35			35			
Fuel Tank Capacity (gal.)	20		20			20			20			
Cargo area												
Inside Height (in.)	62½		66½			72			61½			
Inside Width (in.)	63½		68			75			69¼			
Width Between Wheelhouses (in.)	48		48			48		41	40			
Floor Length (in.)	85 1/8	93 1/8	93 1/8	122 1/8	146 1/8	122 1/8	146 1/8	182 1/8	83½			
Floor Height (in.)	23¼		24			29½		31½	31	31¾		
Capacity (cubic feet)	190	207	232	315	378	375	450	560	See Below			
Rear Door												
Width	36		36			36			37½			
Height	56		60			64¼			55½			
Rear Axle Ratio	5.87:1		5.87:1		4.88:1	5.87:1			5.87:1			
Rear Axle Capacity (lb.)	5,000		5,000			5,500		7,500	5,500	7500		
Rear Spring Capacity (lb.)	1,950		1,950			2,400		3,800	2,500	2500	2800	3800
Front Axle Capacity (lb.)	3,500		3,500			4,500		4,500	4,000	4500		
Front Spring Capacity (lb.)	1,350		1,350			1,750		2,200	1,750			2200
Brakes, front & rear (in.)	12 1/8 x 2		12 1/8 x 2			12 1/8 x 2 front 13 x 2½ rear			12 1/8 x 2 front 13 x 2½ rear			
Tires	6.70 x 15, 6-ply		7.00 x 15, 6-ply	7.00 x 15, 8-ply	7.50 x 15, 8-ply	7.50 x 15, 8-ply		7.00 x 17, 8-ply, doubles at rear	7.50 x 15, 8-ply	7.50 x 17, 8-ply	8.25 x 17, 10-ply (Single)	8.25 x 17, 10-ply (Double)
Turning Radius (ft.)	16¾	17¾	17¾	20¼	22	21	23 1/3	25 1/3	17¾	18 2/3		
							Square Quart Bottle Case Capacity		40	60	80	94
							Optional Refrigeration Unit		Not Available	Available		

fuel economy and reduced maintenance costs of the Perkins engine. Numerous photographs of White PDQ's appear elsewhere on these pages. Of particular interest are photos of power and front suspension modules built around Willys, Continental and Perkins engines and a White PDQ "body drop" over the first power and front suspension module equipped with a Chrysler Slant Six engine, photographed within Montpelier's manufacturing facility.

New orders remained steady and reassuring in 1963. A contract for 220 forward control Urbans mounted on Ford chassis for the U.S. government was received in April. In early May REA and Nickles Bakery placed repeat orders valued at over a half-million dollars. September brought the largest order to date for PDQ's as the Wanzer Dairy of Chicago elected to update its fleet with one hundred insulated and refrigerated PDQ models.

After serving the division as its director of planning, Harold "Buzz" Bartick was promoted late in 1963 to the position of Operations Manager. Bartick had joined Montpelier Manufacturing in 1954 to jump-start the company's development of fiberglass reinforced plastic body panels.

The Montpelier Division's fortunes were bolstered in 1963 with the introduction of new fiberglass and aluminum composite bodies mounted on forward control chassis supplied primarily by Ford, Dodge, Chevrolet and GMC. In appearance and design, these trend-setting trucks rendered virtually all other forward control trucks obsolete. The trucks were so advanced the average person would be hard pressed to discern a 1963 from a 2003 model. The new state-of-the-art trucks were a cooperative design effort between Glen Taylor, Montpelier Division's Chief Engineer, and Walter Chanda, vehicle manager for Continental Baking. Robert Boszor, Montpelier's experienced pattern shop foreman, was instrumental in establishing the feasibility of manufacturing a body with the weight saving combination of aluminum and fiberglass reinforced plastic. In pointing out features of a typical body intended for delivery of baked goods, John Rundell emphasized that, "All structural parts for the Continental (Baking Company) truck body, for example, are aluminum. Wherever there is any place susceptible to rust or corrosion it is made of fiberglass. This combination is better than using 100 per cent fiberglass."

Continental Baking took initial delivery of fifty-one of these new trucks with follow-up orders for 150 more. The first of these advanced bodies were mounted on a Ford chassis and most were powered by a four-cylinder Perkins 4-236 diesel engine. With Continental Baking's total fleet size approaching nearly 14,000 trucks, the joint development effort between the two companies gave the Montpelier Division almost exclusive access to one of the largest and most desirable customers in the company's thirty-seven year history.

Expanded use of aluminum for the new delivery bodies may have come, at least in part, at the urging of White Motor Company. White and its competitors had been using aluminum bodies and frames for lightweight versions of their heavy-duty on-highway semi-tractors for more than ten years. White may have

This Diamond T Trend box truck is equipped with a steel-framed plastic cab developed by the Montpelier Division of White Motor Company. Identical cabs were also utilized for Trend trucks produced by the REO and White Truck Divisions. A gasoline fueled 327 CID Chevrolet V-8 provided motive power for all Trend trucks, regardless of the particular truck brand.

dictated more extensive use of aluminum, a material that the field was already accustomed to, to impart greater confidence in truck longevity that fiberglass by itself could not do. The most profound influence for choosing aluminum as the primary body material probably came from the J.B.E. Olson Corporation of Garden City, New York. Olson "Kurb Side" bodies had been in production since 1947 and, by the early 1960's, were commanding a significant share of the market for forward control delivery truck bodies.

In January 1964, John Rundell spent two days in management meetings at White's Cleveland headquarters, taking several opportunities to enlighten company-wide executives about "the Montpelier story" emphasizing the significant growth achieved by the division over the preceding two years. The division's performance prompted H. J. Nave, White's executive vice-president to enthuse that, "…the Montpelier operation will be, in proportion to its size, one of the outstanding divisions of our company in 1964."

The Montpelier truck body designed in conjunction with Continental Baking was just too good for White Motor Company to pass up. White insisted upon having its own version of the truck and by the spring of 1965, it was released as the White Aluminum PDQ. A full-page ad for the Aluminum PDQ appeared in the May 1965 issue of Commercial Car Journal and is reproduced elsewhere on these pages. Aluminum PDQ customers could choose from any one of four engines, Chrysler's 225ci six and 318ci V-8 gasoline engines were complemented by Perkins in-line four and six-cylinder diesels

Montpelier's order backlog increased to well over one million dollars by mid-February 1964. The division received orders for nearly one thousand units in the first 45 days of the new year, primarily for various branches of the armed forces. The Army and Navy paced orders for more than 300 van type bodies each. The Army, Navy, and Air Force ordered 325 field ambulances while 205 Urban delivery vans

Body panels for Trend truck cabs were molded from sheets of Royalex plastic supplied by the Uniroyal Corporation. The body panels consisted of a tough high-density outer skin that encapsulated a much thicker low-density core. White, REO and Diamond T Trend trucks featuring these lightweight cabs were introduced in 1967.

were ordered collectively by the Army, Navy, and Marines and 150 more aluminum and fiberglass bodied Urbans were scheduled to be produced for the Continental Baking Company. The steady influx of orders called for plans to hire seventy-five additional production workers.

A *Bryan Times* feature article of February 16, 1964 pointed out that sales reached an all time high of $3 million in 1963 and January 1964 sales were up over 10 percent from the previous January. Annual production of both White PDQ and the standard line of Montpelier-bodied trucks was approximately 1200 units in 1963.

The majority of the plant's revenue continued to be generated by Montpelier's traditional product lines; its Urban forward control and its Side-Aisle bodies mounted on cowl and chassis units. Production volumes for White's PDQ line of trucks ebbed and flowed as various orders were filled, but was no more than one-third the volume of Montpelier's traditional products.

In an effort to get more of the Montpelier Division's resources focused on White-branded products and make greater use of its plastic molding expertise, White corporate headquarters commissioned the Montpelier Division to design a new truck cab from Royalex, a plastic material recently introduced by the Uniroyal Corporation that was initially used in the manufacture of canoes. Royalex was a thick walled "sandwich" material with a low-density substrate beneath a tough high-density thermoplastic skin. Starting with flat sheets, the Montpelier Division molded this material into a few large body panels that combined to form a cab-forward cab reinforced by a simple network of steel channels that resembled guardrails. Two prototypes were built in a closed off garage area close to Montpelier's catch up and final inspection department. The development area was off-limits to all but a few employees due to the proprietary nature of the project. The cabs were mounted to a straight rail chassis that would accommodate a box van or a stake platform.

Glen Taylor and the small band of engineers at the Montpelier Division had all they could do to maintain their focus on engineering work related to existing products, so three engineers from the Aro Corporation in nearby Bryan, Ohio were contracted to work evenings at the Montpelier Division to engineer the new truck cab. Leading the small Aro contingent was Dudley Hilliard, Montpelier Division General Manager John Rundell's son-in-law. Paul Henry, Phil Hubbard, and Fred Strobel rounded out the four man engineering crew. Strobel did much of the drafting layout work, drawing the truck cab and chassis at 1/2 scale. At fifteen feet in length, the layout drawing was far longer than the largest of Montpelier's drafting boards, so the drawing was rolled up at both ends as Fred worked on a specific area. Fred recalls that, in addition to their day-to-day engineering responsibilities at Aro, he and each of his colleagues spent an additional thirty hours a week at the Montpelier Division, working evenings on the truck design project.

Although not involved with the truck's development, Montpelier's Assistant Sales Manager Lyle Shoup recalls, "When our prototype unit was built it was to be driven to Laredo, Texas to a test track. I thought this would be a neat way to get a paid vacation trip, so I volunteered to drive it down on my vacation. However the truck wasn't ready to go when my vacation came up so I didn't get to go. I guess it was a blessing in disguise as on the trip down the unit broke down and the driver was stranded in a small town in Texas for a week waiting for repair parts." The ultimate outcome of the test that was intended to be performed in Texas isn't known, but the prototype truck was driven for thousands of miles on back roads in Williams County, Ohio by Leonard Klein, never more than a twenty minute drive from Montpelier.

The truck entered production in the 1967 model year. Oliver, Diamond T, and REO dealers each received versions of this truck that differed only by the brand identification badges. Marketed as the "Trend" truck, its only available engine was a 327ci Chevrolet gasoline V-8. Although the Montpelier Division was responsible for the truck's development and hoped one day to be selected as the manufacturing site for it, White chose, somewhat ominously, to produce the trucks elsewhere. White elected to sell the White branded Trend trucks through its Oliver agricultural equipment dealer network instead of through White truck dealerships. White hoped that selling the truck through Oliver dealers would open up a new market for them in various light to medium duty agricultural trucking applications. White dealers already offered the

White Compact truck that was similar in proportion to the Trend truck. The Compact was a more costly truck with a steel cab and features intended for medium to heavy-duty truck applications that had traditionally been a strong market for White. A Diamond T Trend truck as portrayed in a 1967 sales brochure is shown elsewhere on these pages.

Immediately after becoming the Montpelier Division's General Manager, John Rundell began addressing the financial hardships brought on by White Motor Company's preoccupation with the White PDQ product line and the losses sustained in Montpelier's traditional body-on-chassis truck business. Losses in the previously profitable traditional Urban and Side-Aisle truck bodies were a result of the Montpelier Division's forced commitment to expanding the White PDQ product line. White certainly wasn't pleased with the Montpelier Division's financial losses but seemed unaware that its emphasis on the PDQ line was the driving those losses. Rundell, with his desperately needed Montpelier-centered perspective on the crisis, proposed a much more fiscally prudent course in frank, almost confrontational, discussions with H. J. (Hank) Nave, Executive Vice President of White Motor Company. In fact, the arguments Rundell put forth in his conversations with Mr. Nave were so irrefutable that Nave was persuaded to designate Rundell as successor to Roy Sjoberg as the Montpelier Division's new general manager.

White had been considering the transfer of some or all of PDQ production to another body builder. Given White's disinterest in Low-Step Urbans and Side-Aisle delivery trucks, this move would have emaciated the Montpelier Division and forced at least a temporary, if not permanent, plant closure. With such a decision still up in the air, Rundell, with Nave's consent, began approaching financial institutions with the apparent aim of obtaining financing for an employee buyback initiative.

The following letter from John Rundell to the Central National Bank refers to three enclosures that accompanied the letter; a cash projection statement, a statement revealing the Montpelier Division's cash flow, and a resume of Montpelier Manufacturing. The cash flow and cash projection statements were undoubtedly requested by the bank to evaluate the Montpelier Division's eligibility for a sizable loan. The August 31 letter and accompanying resume are the only documents that are known to survive today and provide, in his own words, Rundell's explanation of the Montpelier Division's problems and his proposed path to daylight and financial strength.

August 31, 1962

Central National Bank of Cleveland
123 W. Prospect
Cleveland, Ohio

Attention: Mr. G. Carlton Hill, Jr.

Dear Mr. Hill:

You will find enclosed as much of the information as we were able to put together in such a short time. We were not able to get the balance sheets which you requested but you will find the projection, the cash flow, and the resume. You realize of course, that the information enclosed is strictly confidential. However, Mr. H. J. Nave of the White Motor Company is familiar with the fact that you and I talked about this proposal yesterday.

If there is anything else which is necessary, please advise me and I will get it to you as soon as possible.

Sincerely,

John C. Rundell

JCR:ph

'his diaper delivery company was based n Evanston, Illinois and may have had outes covering a large part of Chicago. t's not certain if the Montpelier Division vas involved with the paint treatment or these trucks or if the enhancements vere done after the customer took posession of the trucks. This truck and imilar PDQ's in the Crib Diaper fleet vere powered by a four-cylinder Perkins -236 diesel engine.

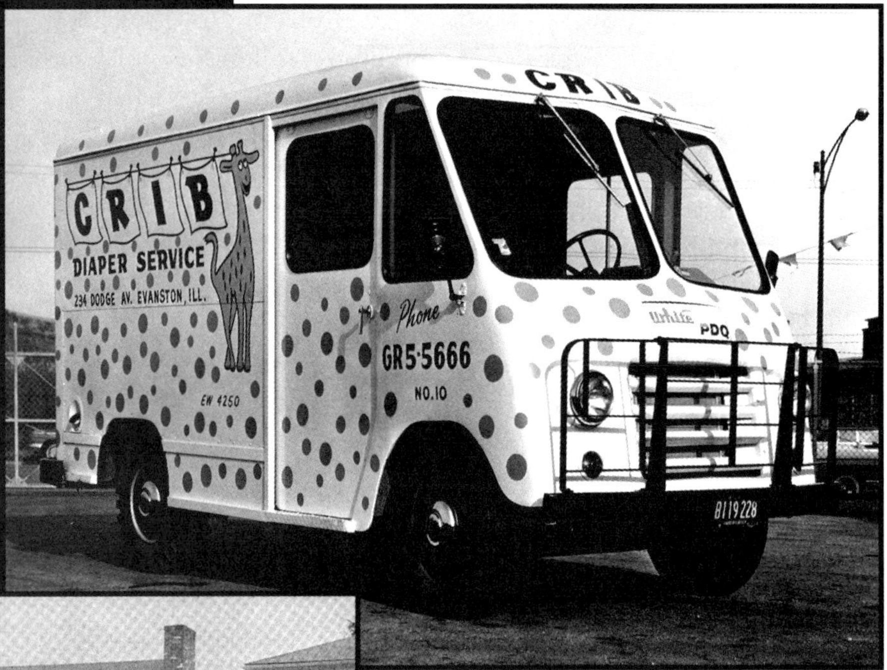

One of 29 PDQ's owned by Merchants' Leasing, Inc., Rochester, N. Y., (a subsidiary of Leaseway Transportation Corp., Cleveland, Ohio) and leased to Rochester Package Delivery.

Here's what 29 PDQ's did for this firm:

Increased average stops per day from 80 to 110.

Increased number of packages delivered per day from 114 to 146.

Increased average miles per gallon from 6.5 to 8.7.

Reduced maintenance costs per mile from 2¢ to 1.5¢.

Reduced size of fleet from 72 to 50 trucks.

Isn't it time you investigated PDQ's?

Call your WHITE distributor or PDQ dealer for a demonstration.

THE WHITE MOTOR COMPANY
PDQ Division · · · Montpelier, Ohio

PICK-UP **P** DELIVER **D** QUICK **Q**

A-297 — This advertisement appears in Commercial Car Journal and Fleet Owner, September, 1961

White's PDQ marketing effort included persuasive fact-filled testimonial ads in trade journals. It's obvious that certain efforts to promote PDQ's were very aggressive but the sales commitment dissipated somewhat at the dealer level. The high cost of producing the PDQ, stiff competition and profit margins that were thin to non-existent kept sales at modest levels

RESUME - Montpelier Division - White Motor Co

The Montpelier Manufacturing Company was formed in 1927 by Mr. H. A. Schwartz. The company was a wholly owned family company with Mr. Schwartz's mother-in-law as President, and his wife as the other principal stockholder.

During the years of 1957 and 1958 practically all of the efforts of the Montpelier Manufacturing Company were pointed toward the development of a unitized vehicle with certain exclusive features that prompted the acquisition of the Montpelier Manufacturing Company, October 1, 1959, by the White Motor Company, Cleveland, Ohio.

We operated as a wholly owned subsidiary of the White Motor Company with Mr. Schwartz as President until June of 1961 at which time he retired and eventually we became a Division of the White Motor Company rather than a subsidiary.

After Mr. Schwartz's retirement on June 1st, Mr. R. H. Sjoberg was appointed General Manager.

At the time of the acquisition there were several errors in judgment made so far as the projected sales potential of the unitized vehicle which had been developed and which became known, and now is, the White P. D. Q. vehicle. These projected sales were originally predicated on a production of 2000 to 2500 units per year. This did not happen, for several reasons, principally the models that were available at the inception were not complete and the White Sales Organization was not familiar with selling light duty trucks as their personnel had been particularly trained in the marketing of heavy duty vehicles. The combination of errors in judgment together with the inability of the White Sales Organization to produce orders resulted in a considerable loss since the acquisition by White Motor Company.

The continued losses were one thing, however, the error in sales projection at the inception constituted an inflated inventory on most of the slower moving models, some of which is still in inventory. This inventory at one time was approximately one million, two hundred and fifty thousand (dollars), and now has been reduced to approximately seven hundred thousand.

The losses which White has sustained at our plant at Montpelier has again been augmented by losses which they have sustained in their branches as quite a number of demonstrator units were put out with the wrong specifications so far as models, etc., which has been sold on a distressed basis, thereby making this P. D. Q. a very unprofitable venture.

Possibly another error that was made, was the fact that much of the success of the original company was predicated on building bodies which were mounted on Chevrolet, Ford, Dodge and GMC Chassis as furnished by the manufacturer, and on which a considerable number of fleet accounts had been developed such as Marshall Field & Company, Jewel Tea Company, Omar Baking Company, Nickles Baking Company, to mention a few. However, at the time of the acquisition and from then on until just recently, practically all of the effort was put on the P. D. Q. vehicle and the body business was relegated to second place. Had the decision been made that we continue to promote the body business as such until the P. D.Q. would have taken over our production, we would not have sustained the losses which we have incurred.

Proof of this; in May of this year the thought was being given by White to close this plant and move certain P.D.Q. production on a contractual basis to another body manufacturer.

However, after talking with Mr. H. J. Nave, executive vice President of White Motor Company, and his faith in my judgment that this operation could be put on a profitable basis by cutting out unnecessary overhead and promoting the body business, I was appointed General Manager on July 1, 1962.

I promised Mr. Nave that I would have the operation in the black within 90 days. We will better this. We only had a little over a $2,000 loss in July with a half months production, and although the figures are not complete as yet, we will show a black figure for the month of August.

To give you some idea of my personal background with the company, I became associated with Mr. Schwartz in 1947 and started Labor Relations inasmuch as he was petitioned by the UAW-CIO on the same day I became associated with him. We lost an election to the union by six votes. However, I was able to renegotiate for a period of one year without getting a contract and/or having a strike and at the end of the year the union filed a disclaimer that they no longer represented a majority of our employees. I then went out into the plant as Plant Manager and from there into sales, first as Fleet Sales Manager, then as General Sales Manager, and eventually into Vice President in Charge of Sales. At the time of the acquisition I was serving in the capacity of Vice President and served this post until the subsidiary became a Division and at that time was made Operations Manager.

I had a very good teacher in Mr. Schwartz, and what success I may have attained in this industry is the result of a lot of patience and training on the part of Mr. Schwartz. I personally have handled such accounts as Bell Telephone Company, Sears Roebuck Company, Railway Express, to mention a few.

In conclusion, I can only restate that I not only am convinced that the potential is above average, especially in view of the potential sales from White each month together with 34 years of know-how in the industry as far as the body business is concerned. According to my auditor we have one of the lowest factory overheads for this size operation that he has ever been connected with.

White
PDC

One significant factor that Rundell apparently decided to filter out of his explanation for the poor sales of the smaller White PDQ A Series trucks was the emergence of small forward control vans manufactured by Detroit's Big Three, the most significant of these being the Ford Econoline van. These low cost, light weight vans offered over two hundred cubic feet of cargo volume, convenient access to cargo from rear doors, side doors, or the driver's compartment, and immediately cut a wide swath through the compact delivery truck market. Sedan delivery trucks and panel trucks, staples in the delivery truck industry for nearly four decades, were suddenly rendered obsolete by these new compact vans. With outstanding maneuverability in traffic, low initial cost and low operating costs, they also encroached heavily on sales of compact forward control delivery trucks like the PDQ and International's Metro-Mite.

Divco produced a compact forward control delivery truck prototype with a unitized body that was almost identical to the smaller White PDQ A Series trucks in 1960. It informed its dealers that it had a major customer-in-waiting for these trucks and asked each of its dealers for input on the truck's marketability to existing and prospective customers. Divco dealers, largely confined to selling trucks for retail delivery of milk and dairy products for many years, weren't the least bit enthused about the potential for expanding their markets with Divco's proposed truck so the it never progressed past the initial prototype.

It's certain that Rundell called the banker's attention to the Montpelier Division's large order backlog as an indication of the confidence Montpelier's customers had in the firm and to impart similar confidence in the minds of prospective lenders. As of August 1, 1962, the division's backlog stood at $650,000. Over 150 units had been ordered within the previous ten days, 72 of which were for the Railway Express Agency, the second order that REA had placed with the company within the previous sixty days. Orders for Side-Aisle utility trucks from Bell Telephone, PDQ trucks for Nickles Bakery and various trucks for the U.S. Army Ordnance Department were also on the books.

Days prior to Rundell's letter, Montpelier's *Leader-Enterprise* published a photo of a dozen Montpelier-built REA trucks loaded on railcars bound for points west, part of an earlier REA order for 147 units. The caption below the photo indicated that twenty PDQ's would soon be bound for Puerto Rico, followed by eighteen PDQ's for a Buffalo customer and thirty-nine refrigerated PDQ's for the H.P. Hood & Sons Dairy in Boston.

Rundell's attempts to obtain financing for an employee buyout of the Montpelier Division were unsuccessful. Left with no other choice, he resorted to his well-honed negotiating skills in a relentless effort to keep White Motor Company's top executives

A 1963 or later model White PDQ C Series truck about to be mated to its engine and front suspension module. A Chrysler Slant 6 engine is installed in the module. This truck is equipped with a translucent FRP top molded only a short distance from the body drop station.

aware of the Montpelier Division's growth potential and its excellent range of products in various stages of development and prototype testing. The two-fold task of encouraging the Montpelier Division's small engineering group to proceed as quickly as possible with new product development while pleading with White Motor Company to abandon plans to close the plant took a heavy toll on his personal health and he suffered a heart attack in the fall of 1964. The citizens of Montpelier breathed a collective sigh of relief as his condition improved and he returned to work, while still recuperating, in mid-November. Not many Montpelierites were aware of the enormous challenges that confronted him until a November 19 editorial in the *Leader Enterprise* brought them to light. Speaking for the local citizenry, the editorial stated, "We're glad to see John Rundell back at his desk. John has done much for Montpelier through the years--through Williams County General Hospital, through his church, and through White Motor Company. When he took over management of White, he put his own ideas to work and the Montpelier Division has been going strong ever since.

"What may not be generally known is that he saved the Montpelier operation for the town. At one time, the Cleveland management seriously considered closing the Montpelier plant, but John had enough confidence in its prospects to fight to keep it here.

"Keep it here he did. And sales and employment have grown rapidly. To their great credit, 'Cleveland' has also recognized John's importance to their business, permitting him to develop the Montpelier Division to its present size and strength."

With a diversified range of White PDQ trucks and Montpelier delivery bodies and the highly capable leadership team of John Rundell and Assistant General Manager Bill Sayers, 1965 was shaping up to be the Montpelier Division's best year ever.

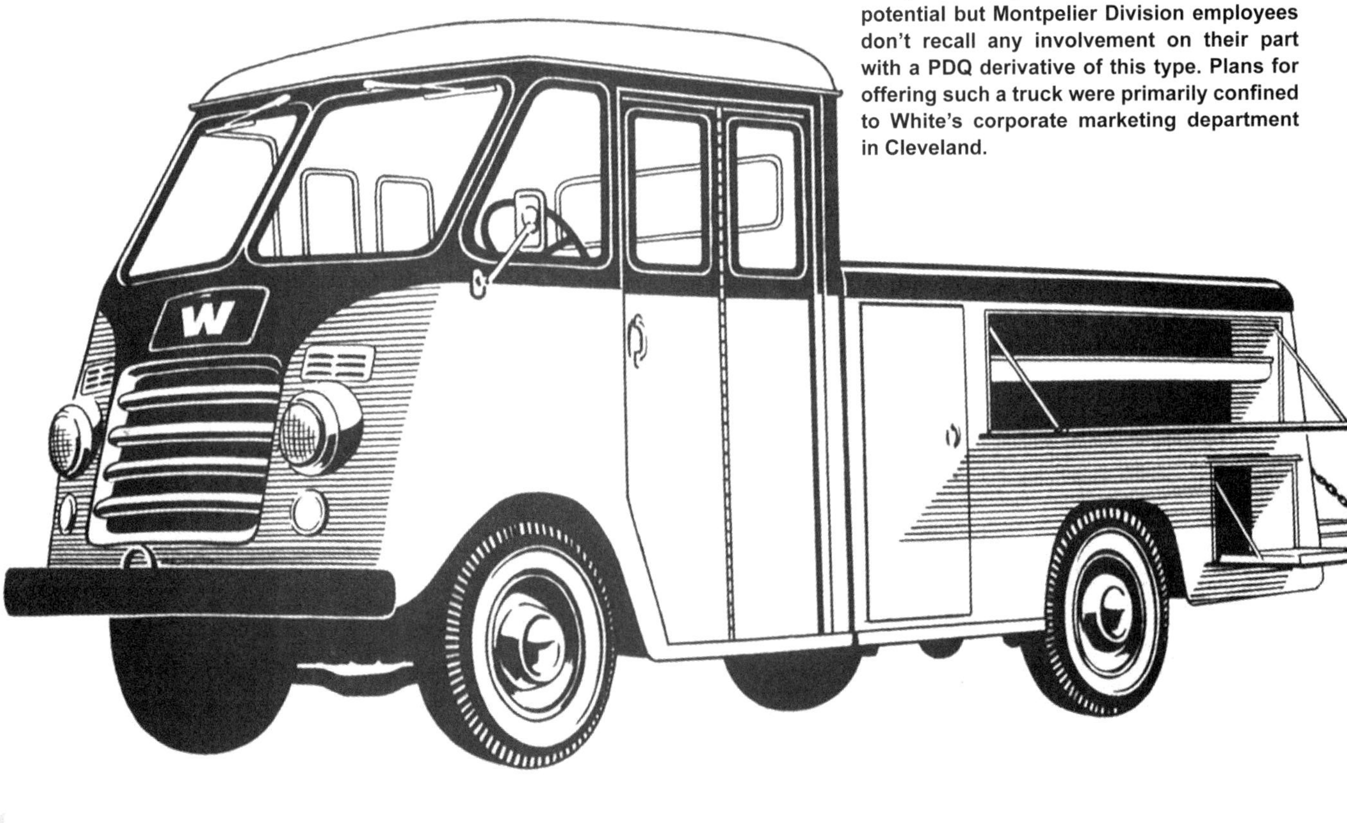

This lineart drawing of a White PDQ configured into a electric or telephone company utility truck may have had some real market potential but Montpelier Division employees don't recall any involvement on their part with a PDQ derivative of this type. Plans for offering such a truck were primarily confined to White's corporate marketing department in Cleveland.

15
Working at Montpelier Manufacturing

The history of a company is formally recorded as balance sheets, financial data, and lists of corporate officers. But the soul of a company rests in the employees who work daily to create the products the company sells. Montpelier Manufacturing was fortunate to have a dedicated workforce that toiled under sometimes difficult and dangerous conditions to produce the truck bodies that made Montpelier successful. Through interviews and historical records, your author has been fortunate enough able to capture some snapshots of what it was like to work at Montpelier Manufacturing.

Shop-Floor Engineering at Montpelier

In a February 16, 1964 *Bryan Times* feature article, General Manager John Rundell praised the Montpelier shop workers, asserting that, "Just because a man is 50 years old he doesn't have to be turned out to pasture. I'm depending on men of this age. Many of our 125 employees have been on the job 20 years. These fellows can do a job without a lot of detailed drawings. They have the know-how. I'm speaking of those people out there in the plant," as he pointed with pride to the entire shop area.

Aerial view of the Montpelier Manufacturing plant taken in November 1946, facing eastward. Note the convenient access to Walbash Railroad rail siding, large quantity of crowl and chassis units awaiting Side-Aisle delivery bodies. Forward control delivery trucks and chassis are relatively few in number. Randolph Street appears in the upper left corner of the photo.

Assistant General Manager Bill Sayers agreed with Rundell, adding that the shop workers "often take something that doesn't exist and put it into tangible form in sales and production. Yes, lots of good ideas come from our veteran employees and, when coupled with teamwork, account for our success."

One of these men was Bill Kunsman. Bill was hired after successfully interviewing with Plant Superintendent Sprankell in August 1936. Having done a fair amount of wiring in his previous jobs, Bill was given the task of making the necessary modifications to wiring harnesses that arrived with Chevrolet, Dodge, Mack, and International chassis and installing the modified harnesses in Montpelier delivery truck bodies before the bodies were mounted to the chassis. It wasn't long before Bill was also installing modified gas tanks in side-aisle delivery trucks destined for dairies, bakeries and newspaper publishers.

With the business changes that were brought on by World War II, the company needed additional welders and Bill volunteered for training in arc and acetylene welding offered by Bob Boszor. His newly acquired welding skills were critical to the war production effort and, as a result, the draft board granted him a deferment from military service. Starting with panels shaped with incredible deft by Morrie Hummel and Lee Henry using electric hammers,

A list of Montpelier Division employees with ten or more years of service to the company, provided to White Motor Company at their request. Some seniority dates conflict with employee recollections and other sources of information. Nevertheless, this is the list that was forwarded to White corporate headquarters.

Employees with Ten or More Years of Service Montpelier Division					
35 Years or More		Date of Hire	15 Years or More		Date of Hire
William Sprankell	37 years	9-10-1927	Norman Mercer	18 years	3-26-1946
Ferman Haase	36 years	11-26-1928	Clair Roush	18 years	9-24-1946
			Paul Strawser	18 years	4-23-1946
25 Years or More			Denny Miller	18 years	9-25-1946
George Cornell	25 years	9-1-1939			
Howard Hiner	27 years	8-16-1937	10 Years or More		
Paul Young	25 years	12-9-1939	Elmer Clark	14 years	4-15-1950
Wayne Bordner	28 years	4-10-1936	Leo Mansfield	14 years	10-18-1950
Marvin Hiner	29 years	6-1-1935	Eldeen Esterline	12 years	7-28-1952
Verlyn Heller	27 years	3-11-1937	Walter Graves	14 years	7-31-1950
Robert Boszor	26 years	7-11-1938	Lewis Lautermilch	12 years	10-6-1952
Robert Treer	25 years	3-20-1939	Harold Bartick	10 years	7-19-1954
William Freese	27 years	8-5-1937	Jumar Kissinger	13 years	2-2-1951
			Ray Rigelman	14 years	10-30-1950
20 Years or More			Paul Altman	10 years	12-13-1954
Martin Holt	22 years	9-22-1942	Blan Esterline	11 years	12-20-1953
Max Kunkle	22 years	1-1-1943	Henry Pape	10 years	1954
Leland Haines	21 years	7-20-1943	Tom March	10 years	1954
Glen Barrett	21 years	12-31-1942			
Herman Manley	22 years	9-11-1942			
Leland Smith	21 years	12-28-1942			
Arthur Donaldson	21 years	7-12-1943			
Jay Ransom	22 years	12-7-1943			*May 8, 1964*

Bill welded the panels together to form complete fenders, front ends and roof front sections. The proof of a weld's quality was soon revealed after front end panels were returned to the panel fabricating shop to have the welds hammered flush. Good welds penetrated well into the mating steel panels. They flattened in compliance with the hammer's repeated blows and, after being hammered out, were barely discernible to the naked eye. Bad welds failed to penetrate the metal sheets they had supposedly joined, fracturing as they resisted the hammer's coercive impacts, consigning the entire workpiece to the scrap pile.

Bill was one of several men on the shop floor that Harry Schwartz would occasionally approach with what he called, "an engineering assignment." Although he lacked the formal training that Montpelier's engineers had acquired, Schwartz knew that he could count on Bill and selected other skilled men on the shop floor to get the ball rolling on certain projects faster than the more deliberate and formal approach taken by the company's engineering cadre. In the mid-1950's, Bill was selected to be an engineering apprentice assisting in the development of the Mono-Lite Model 75.

Working the Shop Floor

Al Goebel began working at Montpelier Manufacturing in 1958 as a painter of smaller sheet metal components shortly after the new Low Step Urban models were introduced. Smaller parts and sub-assemblies were first cleaned with rags and paint thinner, then primed and set aside to dry. Finish coats were applied in a second paint room. There was no forced ventilation in the area where these smaller parts were cleaned and primed and dizziness or intoxication set in after a few hours of inhaling volatile fumes from the paint thinner. Painting smaller parts netted Al 20 cents each while larger sheet metal parts or welded assemblies paid as much as 45 cents each. Painters wore a hood (actually a glorified grocery sack with a drawstring that was cinched around the painter's neck that concealed all but the painter's eyes) with a hose that supplied shop air at low pressure that evacuated paint fumes and provided relatively clean air for breathing. The air supply for the hood was teed to the hose that supplied compressed air to the paint spray gun.

After several months, Al moved from painting smaller parts to painting and priming complete truck bodies before they were mounted on chassis. Along with Ed Miller, Al cleaned all interior and exterior body surfaces with rags soaked with paint thinner before priming the bodies and allowing them to dry overnight. After sanding, primer dust was removed with a blast from compressed air blowguns, followed by a thorough wiping with tack rags. Door and window openings were masked off and interior surfaces of the bodies were painted either silver or light gray and allowed to air-dry overnight. The following morning, door and window openings would be masked off again to protect the freshly painted interiors while the exterior body surfaces were painted. Two-tone exterior paint schemes were available and required roughly twice as much masking and painting time as single color exteriors. Hinged rear doors and all side doors were primed and painted separately and installed after the bodies were mounted on chassis. Roll-up rear doors, employed on REA truck bodies and certain side-aisle bodies, were installed prior to the painting operation.

After exterior painting, the bodies would be rolled on carts into the "oven" (more precisely, a large partially insulated room with several gas-fired heaters) where the exterior and interior finishes were cured at elevated temperature for about three hours. Once the paint was cured and the oven doors opened, Al would slither under the bodies on a creeper to undercoat them while they were still in the oven. Undercoating was by far the most untidy of all the manufacturing processes within the plant and, at the end of each day, Al and his clothes bore a layer of undercoating that was nearly as thick as the undercoating he applied to Montpelier's truck bodies. Al and others in the paint department would don their paint clothes at the beginning of a shift and by shift's end, they would scrape the paint build-up from their shoes with a putty knife. Painters were responsible for furnishing their own work clothes; they would remove them before leaving work and would finally discard them after about a month's usage. Before starting work each day painters would cinch their pant cuffs tightly so that overspray wouldn't work its way up their pant legs. They also greased up their faces at the beginning of a shift so that the paint would come off easily later on.

The same primers and paints were used for both steel and FRP bodies. Finish coats on full FRP bodies and steel bodies with FRP tops were not cured in the elevated temperature room but were instead air-dried to avoid damage to the FRP body panels. Most customers took advantage of the translucent feature of FRP tops

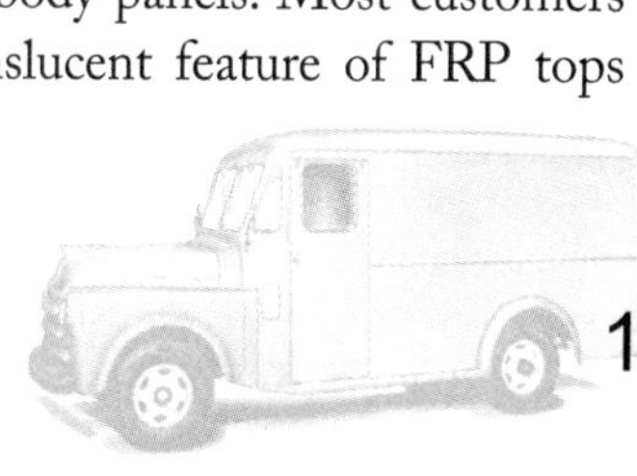

A rear view of a Chicago Tribune Side-Aisle Delivery body installed on a Ford cowl and chassis. The Tribune ordered these trucks without rear doors for newspaper distribution. Bundles of daily newspapers were tossed out the large rear opening while the truck was in transit. A partition protected the driver from nature's elements but this co-worker in the cargo area wasn't as fortunate.

and specified which areas of the top were to be left unpainted to allow natural light to illuminate the cargo compartment. After the paint was fully cured, it would be lightly dry sanded. Decals specific to the customer who was purchasing the truck would be applied and then the entire exterior would be sprayed with a clear coat.

Initially, Al worked the night shift in the paint department and, since he was limited to completing four truck bodies per shift due to restrictions on piece rate reimbursement, he would quickly paint two trucks before his lunch break and then climb on the top of a truck and take a nap until lunchtime. He would then complete his nightly truck painting quota upon returning from lunch. The absence of a night foreman enabled Al to take such liberties with full impunity.

One day as Al was on his creeper undercoating a body, Harry Schwartz and three other "white collar guys" entered the oven area. The timing of their visit couldn't have been worse as Al unwittingly "undercoated" Schwartz and his guests from the knees down. Led by Schwartz, the four men stormed out of the oven room, each of them ranting and roaring as they traversed the shop floor en route to Schwartz's office where Harry continued his lengthy apology while making immediate arrangements for a change of clothes for each of them and himself. When asked by his coworkers if he was aware of the fact that the foursome was in the vicinity of his undercoating gun, Al replied, "No!" but the facetious grin on his face told a completely different story.

Al enjoyed the work that he did and the many friendships of his coworkers, some of whom he had known all of his life. The work wasn't without its frustrations but Al and all of the seasoned employees on the shop floor realized that it was best not to dwell on them for too long. Along with Keith Miller, Al felt that too many processes were jury-rigged and that the aging facilities hampered productivity and made work more difficult. Those two factors may have contributed to the indifferent attitude many shop workers had with regard to keeping their workstations neat and orderly. With rare exceptions, production tooling was not what one might call state of the art. Some workers thought that it was best described as the product of the common man's ingenuity. In spite of the drawbacks, Al consoled himself with the fact that at least he was able to get enough work done to reach his daily piece rate compensation limit.

If scaffolding wasn't available, painters used self-supporting wooden ladders to paint the tops of trucks, getting up and down the ladders and moving them along the length of the truck body as needed to paint the entire top. Al would grab the bottom of a wooden ladder while Larry Bauer was near the top of it and put a scare into him. Larry would reciprocate at another time when Al was least expecting it. They agreed to curtail their antics somewhat after Larry fell eight feet from the top of the ladder that Al lifted up farther than he should have.

Toward the end of each month, the shop would get behind on their truck production schedule. Trucks that had been promised to be available for delivery in a few days hadn't even been started yet. Management would compel and coax employees to make up for lost time and workers would comply as best they could, completing work that was far and above their $20 daily compensation limit for piece work. The tickets for work performed in excess of the $20 daily limit had to be held back or "banked" for a future slow day, if and when one came along. Al recalls that in July 1965 he had $600 worth of banked tickets for work that he had completed but was not yet able submit in exchange for wages.

If bodies weren't properly sanded after priming, the painters would get blamed for a poor quality paint job. Sanders would decline to do parts of the job over again if they had already done the job once, particularly if a foreman had already signed a slip indicating that the work had been completed. Painters would have to resand the bodies before they started painting. Problems of this scope usually couldn't be resolved between workers or foremen so plant superintendents Henry Pape or Denver Miller would be called upon to resolve any disagreements.

With the aid of chain-fall hoists and hooks, painted bodies were lifted off the carts that had shuttled them sequentially through welding, soldering, solder dressing, cleaning, priming, sanding and painting. Chassis were driven through the large overhead door on the plant's south side into the approximate position for the body drop. Paul Houk recalls that measurements were first taken to insure that the bodies were correctly located on the chassis. Four holes were then drilled through the body and the frame rails on each side of the truck and the body was fastened to the frame with ribbed-neck bolts. On Side-Aisle deliveries, several holes were also drilled through pre-punched holes in the body firewall and into the mating chassis cowl. Bolts were then inserted and tightened to secure the firewall to the cowl.

Paul started work at Montpelier Manufacturing in May 1957 and spent the better part of eight years working in the mounting department. By the time Paul arrived, all but a few bodies had steel flooring. Tongue-and-groove wooden planks were an available option for certain body designs. Once the truck body was secured to its chassis, doors, windows, wiring, windshield wipers, interior and exterior lighting, mirrors, a swivel-and-tilt driver's seat, and other accessory items were installed. Heavily insulated trucks destined for hauling dairy products usually were equipped with tracks on which standard sized dairy cases could be loaded. Large cooling coils, also known as holdover coils, were installed in trucks that would be transporting refrigerated or frozen items. Stationary refrigerant compressors at a warehouse or loading dock would refrigerate the coils and the cargo compartments overnight. Other refrigerated trucks were equipped with on-board compressors to continually cool the trucks. Bakeries could select from a wide variety of rack designs that Montpelier offered.

After the trucks were finished and briefly road tested, each truck would go through the "catch up" department at the west end of the plant for cleaning, touch up, and defect correction.

Each employee's work was recorded on pink slips. At the end of a shift Chuck Brigle organized all of the pink tickets according to truck serial number for piecework that was completed and delivered the tickets to the office staff for payroll and general record keeping. Company rules with regard to completed work tickets were strict and inviolate. Al Goebel recalls that a ticket that was last in the possession of one of the yard workers was unaccounted for and apparently windblown. Three or four workers searched the yard for the ticket for an entire day, finally locating it underneath a pile of scrap.

This bakery truck has an optional roof rack mounted to the truck's all steel roof. A ladder hinged to the right rear door provided convenient access to the roof rack cargo and nested against the rear door when the door was closed. The same rack and ladder arrangement was an option on International Metro bakery trucks as early as 1951. Roof racks like the one installed on this truck were sometimes used to transport tires or bags of soiled laundry in other delivery vocations.

Henry Pape joined Montpelier Manufacturing in 1955, working in the yard, moving chassis, unloading incoming material from railcars and loading finished trucks onto railcars. He spent a total of ten years at Montpelier Manufacturing, divided nearly equally between the Schwartz and White Motor Company eras. Albert Meiser, who would soon be retiring, trained Henry in the art of making some of the simpler panels that comprised steel tops for the full range of Montpelier bodies.

A steel top consisted of six or eight panels depending on its length. Corner panels with their compound curves were hammered out on one or more of three electrically operated hammers in Montpelier's sheet metal department. Starting with flat sheet metal blanks, Lee Henry operated the hammers, deftly shaping the domed contours on each of the more complex panels. Henry Pape would then attach a complete set of panels over the top's frame by spot welding with an arc welder. Holes roughly a quarter-inch in diameter in the top panels were aligned with the top's frame and arc welded to the frame in those locations. After the welds were "dressed" with a disc grinder, Leo Mansfield would spread solder over the welds and the seams between the panels.

Welding tops to the top frames wasn't actually Henry's job but he did it and received piece rate pay for it. Henry explained that the worker that usually welded the tops was often "busy up on the line welding so I would weld them myself."

Henry worked with Albert Meiser for roughly a year before Albert retired. Each man worked on a top from start to finish, completing up to three tops per day. Even though they were capable of producing more, they refrained from doing so, knowing that the piece rate for fabricating a top would be reduced if they "overproduced".

Lee Henry was Montpelier's only day-to-day hammer operator and caught wind of an effort to reduce the piece rates for his hammered top and body panels. Lee took issue with the impending rate cuts by removing the dies he made for the hammers and quitting. Montpelier's management could find no one with Lee's indispensable skills to replace him but Lee had produced a large bank of panels (for which he had not yet received compensation) that gave the company considerable breathing room before his replacement could be found. Henry Pape tried his hand at the hammer once or twice but had absolutely no success.

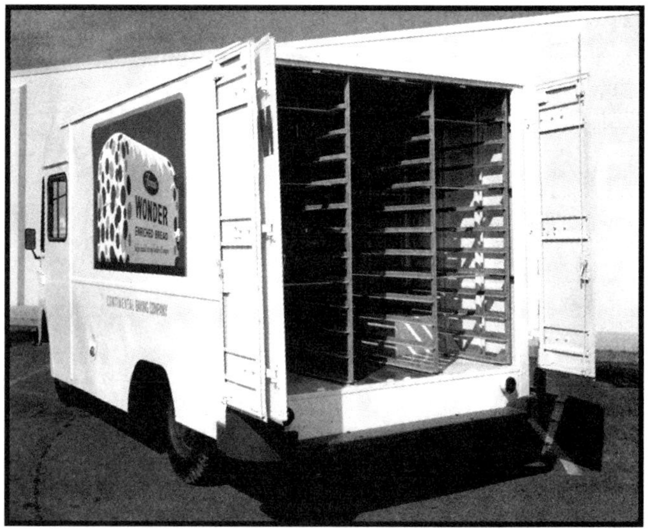

The racks in this truck were precisely matched to Continental Baking Company's baked goods and trays. Note the bi-fold rear doors that provide full width access to the cargo compartment. The doors could also be folded flat against the sides of the truck, enabling the truck to back up until its rear bumper contacted a loading dock, making for smooth loading and unloading of cargo.

After he was promoted to a foreman's position, Henry reported to Ferman Haase, superintendent of the welding and body assembly department. Henry got along reasonably well with Ferman but more than a few shop workers found Ferman's management methods rather annoying. Henry recalls that "Ferman was smart and hard working but he was always hiding and watching you and that just irked the daylights out of us." Ferman later transitioned from the superintendent's job to a position as a manufacturing engineer. Several succeeding superintendents failed to be as effective as he was. According to Henry, "Superintendents were a dime a dozen. If they didn't produce they were gone."

Henry snickers when he recalls that "Harry Schwartz had a knack for walking down the aisle where we built the front ends. Every once in a while he'd pull me aside and say, 'Henry, you'd better pull that front end off the line because it's out-of-square.'

"How do you know that, Harry?" Henry asked.

"I just know." Harry replied and he never revealed to Henry how he was able to make that assessment. Finally Henry realized how Schwartz was able to make his determination. When viewed in just the correct line of sight, a visual check for parallelism between an upper shelf and a length of channel steel on the inside of the cab could be made. If they didn't line up visually, the front end was, in fact, out of square. Henry's welders never did learn how to make that subtle observation and were frustrated by the infallible nature of the Schwartz-Pape visual inspection technique.

Henry set high standards for quality and productivity in the welding department and his good work earned him a promotion to superintendent over the pattern shop and sheet metal and welding departments.

Not long after Henry became superintendent, John, his ten-year-old son experienced what in simple terms could be called a kidney stone "blowout". He was taken to the Williams County General Hospital in Montpelier. Harry went to visit his son on a winter Sunday morning and was alarmed to find that John's condition had deteriorated severely. Henry immediately summoned a doctor and within a short time John was in an ambulance on his way to a larger hospital in Ann Arbor, Michigan. John went through a gruesome and lengthy surgery, very nearly cut in two by an incision that gave the doctors full access to his intestines in order to remove the toxic waste and treat the young lad for peritonitis. Henry recalls that "They lost him a couple of times but they got him back. We had a rough time. The only reason I'm telling you this is to tell you that the whole factory stopped production, everyone came to the front of the factory and they all knelt down and prayed for my son."

An incredibly large order for Railway Express Agency trucks necessitated ramping up production from three trucks a day to eighteen in 1963. These trucks were derivatives of the truck that was jointly designed by Montpelier and Continental Baking, the major exception being that REA specified steel and fiberglass construction in lieu of the lighter weight aluminum and fiberglass combination that was exclusive to Continental Baking. An escalating conflict between a few workers was taking its toll on second shift productivity so Henry worked double shifts until the issues were resolved. With a lot of hard work and will power on the part of all shop workers, the six-fold increase in truck output was achieved.

While Henry was superintendent, the Montpelier Division was in constant competition with another manufacturing firm in nearby Bryan, Ohio for welders. If welders at the Bryan factory received a raise, Montpelier Manufacturing's welders would insist on a similar raise or leave. Henry would encourage other workers to obtain the training required to become production welders but if that contingency wasn't fully successful, he'd spend extra hours in the evenings going from farm to farm trying to hire farmers with welding skills.

The standard joke around Montpelier and the surrounding areas was, "If you want any tools, just hire on at White Motors." Henry was constantly buying new tools, e.g., electric grinders, electric drills, electric reversible screwdrivers, etc. One day Henry took note of the fence along the company's property line and the fact that it needed painting. He went to take a closer look at the fence and stepped on something in the tall grass along the fence. Kicking the grass aside, he found a brand new grinder, one of six that he had just purchased. One of Montpelier's employees was also a part time Montpelier policeman. Henry told him what he had found and gave him a special stake out assignment. Whoever tossed the new grinder over by the fence must have found out about the stakeout because three or four days passed and the grinder was never picked up.

Denver Miller quickly approached Henry one morning asking for help. One of their more valued and highly skilled employees was going back to his car and, as he was leaving the plant, an electric chord was dangling from under his arm. It was one of the new electric tools that Henry had just bought. Henry caught up with the fellow and said, "Man, I just can't believe that you would do something like that. It hurts me to have to tell you this but as of right now you don't work here any more and if you ever set foot on this property again, I'll have a warrant sworn out for your arrest. That's how strong I feel about thieves."

"Every once in a while, I'd call the night watchman in and tell him to stand by the gate and have everyone open their lunch pail as they left. A young 'cracker-jack' guy that worked in inspection refused to open his lunch pail. I told the guard, 'If they refuse, just tell me. Don't make a scene.' The next morning Henry, fully apprised of the man's refusal to open his lunch pail, fired the defiant worker on-the-spot. The young man insisted that he had nothing of the company's in his lunch box but Henry was adamant about the firing because the man refused to obey a direct order that was required of each and every employee leaving the plant."

Your Author's dream Job - Delivering Delivery Trucks:

Leo Clark, the enterprising proprietor of Montpelier's Sinclair service station, had contracted with the Kiel Driveaway Company in Cleveland to deliver finished Montpelier bodied delivery trucks to Montpelier customers throughout the country. One of his drivers was Clarence "Junior" Fisher. Junior's happy memories of these trips have lasted a lifetime. He provides the following account of his experiences as though they had occurred only recently:

"We delivered all over the United States. Hundreds of trucks with our heavily insulated ice cream delivery bodies were driven to Frosty Freeze headquarters in Runndymeade, New Jersey. We delivered REA parcel trucks to Boston, Florida, Los Angeles, New Orleans, Illinois or anyplace REA asked us to. Normally, each of us drove a truck while pulling a second one behind us with a tow bar.

"I was out of work at the time. I had worked down at White Motors, welding and soldering the front ends around the windshields. We would alternate working at White Motors and the old Fifty Plus Five Corporation. They had off-seasons that were opposite of each other, so we would move from place to place as we were laid off. When one would have a layoff, I would go to work for the other.

"Then it turned out that those jobs slowed down, but they needed someone to deliver trucks so we delivered trucks for a couple of years. It was a good way to pick up some extra money. All you needed was your chauffeur's license. You could run it as hard or as easy as you wanted to. It depended on how hard you wanted to work and how much money you wanted to make.

"Some guys did it part time, taking trucks to Chicago, six to ten trucks at a time with one guy following in a car for the return trip home. It gave everybody a little spending money and a chance to be with your friends.

"Once in a while we would have a half dozen or so trucks going to one place. We would normally get our buddies together and go as a group. We always had a better time when we did that. However, I was driving by myself a lot of times. I'd deliver to Texas and Tennessee, anywhere and everywhere, really. Most of the time they would just have two units going out to one place which meant just one driver."

"Every truck that was driven was pulling a truck with a tow bar. Most of the drivers hated handling the tow bars because they were heavy, dirty and greasy. Harold Thomas and I each took a couple of trucks to Boston at the end of February and it was snowing like crazy. Two more guys left just after we did. We delivered in Boston and were unsuccessful in getting railroad tickets for our trip west so we jumped an express train to New York City.

"After we arrived in New York, Harold called the bus station, knowing that they had the "hot shot" (*express route*) that traveled the Ohio Turnpike on its way to Chicago and found that a bus was nearly ready to leave. Harold urged the bus company to hold the bus until we arrived at the main bus terminal by taxi and the bus company agreed. Once we arrived at the bus terminal, the bus driver, already upset that his departure was delayed, refused to sell us tickets because we wanted to stow our tow bars in the bus storage compartment."

The tow bars weighed nearly 60 pounds and, although dirty and greasy, they were carried within a fairly tidy canvas bag roughly four feet long. The bus driver thought that Harold's small town demeanor would be no match for his big city swagger but it was only a matter of minutes before he found out just how savvy Harold was. The impasse was resolved just a few seconds after Harold turned on his heel and started toward the bus terminal, fully intending to call the president of Greyhound Bus Lines. Suddenly, two seats became available directly behind the previously obstreperous bus driver and in due time the men were successful in persuading him to drop them off close to home at a location on the Ohio Turnpike known by Montpelierites as "Pigpen Plaza."

The other drivers that were destined for Boston left Montpelier only an hour or two after Harold and Junior but couldn't successfully contend with the snowstorm's onslaught and became stranded long before reaching Boston. They eventually made it back to Montpelier, a full five days after Harold and Junior's return.

A sizable portion of Montpelier Manufacturing output was transported by rail as well.

The two photos on the next page reveals how the trucks were loaded onto flatcars on the factory's south property line.

Montpelier Manufacturing's simple and effective procedure for loading delivery trucks onto rail flatcars. Cement blocks and boards were used under the ramp's tires in varying amounts, depending on the topography of the rail siding at any given location. What a shame that these bright orange Interstate System trucks were denied a color photograph opportunity!

Office Life

Office life at Montpelier Manufacturing was just as interesting as life on the shop floor and just as worthy of mention. After all, it was the function of everyone in the office to strive together to "make the cash register ring."

Marcielle Dick, a native of nearby Metz, Indiana, graduated from Indiana Business College in Fort Wayne in 1940. She learned of a secretarial opening at Montpelier Manufacturing through the college placement office and, after being interviewed by Harry Schwartz, accepted his job offer in the fall of 1940 at the tender age of eighteen.

Marcielle was Mr. Schwartz's private secretary, typing much of his correspondence, primarily form letters that she had to tailor in one way or another to the specific needs of the addressee. Bob Treer provided additional secretarial support to Schwartz, with Marcielle taking dictation and Bob typing much of the more routine correspondence directed toward customers and suppliers. When Schwartz wasn't dictating a letter, he would often process incoming mail, much of which was placed in a basket for Marcielle to accurately file in roughly six separate categories. The office lacked a switchboard at the time Marcielle was hired; she and Bob Treer were responsible for answering the phone and directing calls to the appropriate person on the office staff.

Railway Express Agency box van trucks secured to flatcars on the rail siding adjacent to the Montpelier Division's south property line.

Marcielle readily admits she began looking for a different job after fully realizing what it was like to work for Harry Schwartz. Over sixty years later, she has no difficulty recalling that, "He was a tough taskmaster, even with *(plant superintendents)* Bill Sprankell and Ferman Haase. He wanted everything done just right, better than they do today." She concedes, with a certain amount of reluctance, that if Schwartz hadn't been as forceful, the company might not have remained a viable entity but she quickly reiterates that, "He was a tough taskmaster."

Marcielle recalls that Mr. Schwartz was the company's sales tycoon. He would make occasional trips to Chicago where Montpelier Manufacturing did a large share of its business and return with a fist full of orders for Side-Aisle and Urban delivery trucks bound in particular for Omar Bakery or Marshall Field.

When a new 26 ft. x 82 ft. two story glazed block office was constructed at the east end of the factory in the early 1940's, additional office personnel were hired. Schwartz's office was near the front door on the north end of the new office. Quarters were provided at the extreme southern end of the office for on-site government inspectors that spent a great deal of time overseeing Montpelier's production of war materiel. Colin Cameron and his drafting work were consigned to the basement of the new office building, an area that would eventually become the location for Montpelier's engineering department.

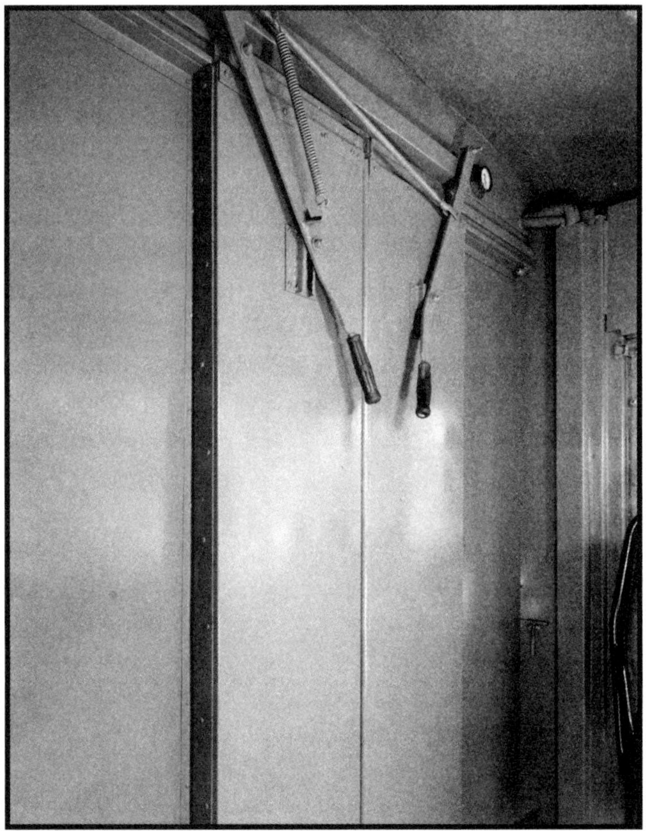

Marcielle was also assigned to the government inspectors for secretarial support. War production sustained Montpelier Manufacturing like no other trade had done before and Harry Schwartz went so far as to provide room and board for Mr. Weeks, one of the on-site inspectors, at his East Washington Street home. During the war, some of the office workers were required to come in to work on alternating Saturdays. Marcielle's usual means of transportation to and from work during the week was not available to her on Saturdays so Mr. Schwartz would pick her up in his Cadillac coupe.

Suspended from the high ceiling in the new office were fluorescent light fixtures, each with two four-foot long fluorescent tubes. Marcielle clearly recalls that, "I had just got done typing and pulled my hands away from the typewriter but hadn't yet pushed the typewriter stand completely away when something

(left) Behold the brightly illuminated cargo compartment of an Urban Delivery body with its translucent fiberglass reinforced plastic top! The absence of exterior paint on this top results in maximum illumination by natural light. Montpelier customers could order their trucks with the FRP top as shown or with portions of the top painted for adequate albeit subdued natural illumination. A partition with sliding door separates the cargo and driver's compartments.

(bottom left) The spring-loaded linkage that opens and closes the doors to this partition positions the spring in an "over-center" arrangement, thus enabling the spring to firmly and securely hold the doors in both the open and closed positions. Either lever can be used to open or close the pair of sliding doors. The heavy-duty partition pressure gage, insulating material applied to the underside of the roof and the piping entering the cargo compartment just below the roof mark this truck as one that is equipped with "hold-over" refrigeration coils.

(below) A view from the opposite side of the partition shown in the previous photo showing the "hold-over" cooling coils installed in the cargo compartment. Lacking an on-board refrigeration unit, the hold-over coils were cooled overnight while the truck was docked at a location that charged the cooling coils, bringing the cargo compartment and cargo to a sufficiently low temperature to prevent spoilage during delivery on the following day. The floor and wheel housings are made from diamond tread plate to minimize slips and falls as the driver maneuvered through the compartment. Note the heavy-duty latch for securing the thickly insulated rear door at the top of the photo.

crashed. I suppose I went white as a sheet and Mr. Schwartz (I can see him to this day, coming out of his office) and Bob ran over to me. One or two of those bulbs had fallen out of the fixture and crashed on my typewriter. That's how close I came to getting my fingers cut off. Mr. Schwartz got his arms around me and had me sit down at another empty desk and he was very kind.

"Another time in the old office I had an awful sore throat but I showed up for work because he didn't like you laying off." Schwartz, able to observe Marcielle's condition quite readily, summoned Bob Treer and said, "Take her up to Dr. Luxan's and get her throat painted."

Under these and other somewhat extraordinary conditions, the softer and more compassionate side of Harry Schwartz would occasionally arise to reveal a genuine concern and sense of compassion for his workers. When he was so moved, it was indelibly imprinted on the minds of those who happened to be around him because it stood in such sharp contrast to his natural work-related disposition.

Marcielle was married in March 1945 and after becoming pregnant in May, she gave Schwartz a month's notice, informing him that she would be leaving in July. Upon learning of her intention to leave, Marcielle recalls to this very day that Schwartz was, "... nasty to me. More nasty to me than any of the times. The last day, he asked me if I would please stay." Schwartz's disgust upon learning of her departure and his more tactful entreaty in asking her to stay were his ways of acknowledging that she was a valued employee and that he would miss her helpful presence in the office. Schwartz's regard for Marcielle was typical of the deeper feelings he had (but rarely revealed) for all employees who faithfully served the company. Marcielle is among a group that was photographed in the old Montpelier office one April afternoon in the old office building, shortly after she was hired. That photo can be seen in Chapter 5.

Kay Cox interviewed for a secretarial job at Montpelier Manufacturing a few weeks before graduating from Montpelier High School in 1959. She accepted Montpelier Manufacturing's offer of employment and, at a wage of $1.00 per hour, began working in the office right after graduation. Her inside connection to the company came through her future mother-in-law who, as the Schwartz's housekeeper, gave her a glowing recommendation. Typing up price sheets, learning to operate the switchboard and serving as the plant's receptionist were her primary duties during her initial sixteen-month stint at the body works. She left in October 1960, got married and joined her husband, a United States Marine, in North Carolina. A year later, she returned to Montpelier and immediately resumed work at Montpelier Manufacturing as a secretary to Sales Manager Ron McCrary and backup secretary to Harry Schwartz. (It wouldn't be inaccurate to say that everyone in the office was a backup employee of Harry's, except for times when he gave you a direct order. At those times, his directives were top priority and all normal duties were subordinated to a backup status).

Versatility on the part of female office employees was the rule rather than the exception. Once Kay had mastered switchboard operation, she was given the task of passing her skills along to a newer female employee. The new girl, still a novice at the switchboard, became a little overwhelmed when she found it necessary to connect three simultaneous incoming calls to three different members of the office staff. With a phone chord in each hand, the new gal's inexperience proved to be rather painful as the third call rang in. To free up a hand to connect the third call, the novice switchboard operator hastily clenched one chord with her teeth. The resultant electrical shock knocked her out of her chair, screaming for help. Kay rescued her understudy and, after restoring order to the switchboard, she and the rest of the office staff had a hearty and protracted laugh at the expense of Kay's utterly shocked and embarrassed trainee.

Kay reluctantly ventured out on the shop floor when necessary to check on inventory levels in the stockroom or to find out if a certain person's time card was correct or if they had forgotten to punch in. High heels were standard footwear for female office workers at the time and the oil-slicked wood floors with all manner of ruts and holes posed a definite safety hazard. The smell of oils and solvents permeated the shop's atmosphere and smartly attired female office workers spent as little time on the shop floor as possible. She found the shop floor be vast (at least in relation to the size of the office) and its production activities intimidating.

On one occasion, however, she ran an errand on the shop floor that proved to be uniquely entertaining. Out of the corner of her eye, she spotted another woman from the office shuffling across the shop floor in a rather awkward fashion. A closer look revealed that the woman was kicking a small object across the floor while trying not to look conspicuous in front of the men working in the paint and final assembly areas. Once she had pushed the object all the way to the east wall of the shop, she carefully stooped over, discretely picked the object up and slipped into the office with it. Brimming with curiosity, Kay returned to the office and prodded the woman to tell her what she was doing out on the shop floor and what the fuss was all about. Finally the woman relented and told her that one of her garter straps had become detached at both ends and had fallen to the shop floor. She realized that she had lost the garter strap only after having walked a considerable distance past it. Rather than going straight back to pick it up, she thought it wiser to nonchalantly circle back to the location of the stray lingerie component and kick it, inch by inch, to a low traffic area on the shop floor before actually picking it up. The men on the shop floor found the loss and retrieval process so interesting that they refrained from disturbing the woman, just to see how she would accomplish the delicate retrieval.

If the office provided a safe haven from the smell of solvents in the shop, it did little to attenuate the odor of ammonia from the blueprint machine in the basement of the office building where the engineering department was located. In addition to producing blueprint copies of original engineering drawings, the foul smelling machine served as the office copier. If enough carbon copies couldn't be produced on a typewriter, the clearest of the carbon copies was run through the blueprint machine. Handwritten notes were occasionally reproduced in the same manner. A ditto machine was acquired in later years, significantly reducing reliance on the blueprint machine by non-engineering personnel. Kay would take brief breaks to chat with the engineers in the office basement and even make coffee for them at times when the ammonia smell wasn't overpowering.

Kay recalls that Harry Schwartz was blunt and overbearing and, with rare exception, he was all business. Without hesitation, he would leave his private office and investigate what he considered to be too much laughing or an excessively lighthearted atmosphere in the common office area shared by the secretarial and administrative pool. Working as a secretary to Sales Manager McCrary and the small sales department, Kay spent nearly ninety percent of her time tediously typing detailed and lengthy price quotations. Each customer-specific option had to be painstakingly typed from a master options list; word processing and the luxury of cutting and pasting standard text was still decades away. Montpelier's sales force was housed within the confines of the Montpelier Manufacturing office. There were no regional sales managers or sales representatives strung out across the country. When sales activity couldn't be handled by phone or written correspondence with or without enclosed brochures, specifications sheets, and price lists, a well-planned series of customer visits was planned for the purpose of closing impending sales and stirring up new business with both new and well-established Montpelier customers.

Shortly after starting to work for McCrary, a sales quote that she had typed was routinely signed by McCrary and mailed to a customer. The customer called Harry Schwartz a few days later, noting that the final one or two paragraphs with important pricing information had been omitted. Schwartz was enraged, stormed out of his office and sternly excoriated Kay to the point of tears in the presence of her fellow workers. Unable to get a satisfactory explanation from Kay as to why McCrary had signed the price quote, McCrary became the next recipient of Schwartz's cutting invectives. Needless to say, Kay and McCrary atoned for their omission in a quick and thorough manner. Although Kay was devastated by Schwartz's verbal torrent, she credits him with a willingness to quickly forget about such unfortunate incidents, *never* holding them as a club over one's head once sincere apologies were offered and reparations were made.

When Schwartz went out onto the shop floor, it was often to investigate a production problem that wasn't being solved with the level of alacrity that he expected. At other times, he'd MBWA (manage by walking around), just to be satisfied that work was being done as expected, i.e., swiftly and correctly. Occasionally, he'd quiz his production workers to see if they were in command of their tasks or to find out if they were encountering production difficulties or being hampered by production bottlenecks. Shop employees never approached him with an issue or problem because they didn't want to ignite a firestorm and, tacitly, they knew that he expected a smart shop worker to work on his own or with his foreman on solving problems.

There were times when shouting matches would occur between Schwartz and various shop floor supervisors that the closed doors of his office were grossly incapable of muffling. Although he was stern and even vituperative, he was also fair and his subordinates could gain his respect if they respected him and didn't step over the line and try to prove themselves smarter than him. He could be convinced in an argument that someone else was right, if presented with good factual information. In the community, he was a very active and caring individual. As an astute citizen, businessman, engineer and business owner, he possessed all of the essential requirements for the positions of authority that he held.

Kay describes her days at Montpelier Manufacturing as excellent business experience, a great training ground where the work ethic was much better then than it is in the 21st Century workplace. Learning how to multi-task, accept a lot of unexpected "walk up and drop off" work, take lots of shorthand and type scores of dictated letters increased her office capabilities and capacities tremendously. Looking back, she feels good about the personal and professional growth she attained at the body works and the way it prepared her for the years to come.

Kay's tenure at Montpelier Manufacturing spanned a period time that was both before and after the White Motor Company purchased the company. She also had the opportunity of working under and observing the contrasting management styles of two general managers. Virtually everyone was grateful for his or her job and glad to be working at the body works, even under the domineering leadership of Harry Schwartz. Employees accepted Schwartz and his sometimes vitriolic temperament because they knew that he would spare no effort in keeping the company (and their jobs) secure and viable. There were few layoffs during the late 1950s and the years that followed. Work was stable, people showed up faithfully and there were no shirkers in the office or on the shop floor. Employees were a good blend of younger employees and older skilled workers

who willingly shared their skills with newcomers. From Kay's perspective, there was no backbiting, no one ran the company down, there were no turf wars and no one felt or acted more important than anyone else throughout the entire organization.

After Montpelier Manufacturing became a division of the White Motor Company in 1959, Kay thinks that a sense of reduced employment security permeated the minds of Montpelier's salaried and office hourly employees. In the minds of the office staff, White was a sprawling corporation with divisions spread out over all of North America with no long-standing ties to little Montpelier, Ohio and not at all dedicated to keeping the business in Montpelier. Turnover within the salaried and office ranks definitely increased after the White buyout with people departing for what they thought would be a more secure future at other businesses in the area.

White made very few procedural changes or upgrades within the factory after the purchase, leaving the division's operation in the capable hands of its founder, Harry Schwartz. Sales methods did change somewhat; typing extensively detailed price quotations was abandoned in favor of less labor-intensive procedures. Electric arc welders on the shop floor were upgraded and White set higher quality control standards for welds. When a White Motor Company entourage arrived from Cleveland for a plant visit, their expensive suits and flashy cars alienated them from many of the common folk working in the office and shop. Harry Schwartz and John Rundell wore suits and drove fancy cars but they were integral parts of the business and still able to relate well with all of the shop and office employees.

The sole reason White bought Montpelier Manufacturing was to have exclusive access to the White PDQ line of delivery trucks. Kay recalls that "PDQ" was the "buzz word" in the office and shop. White made sure that the Montpelier Division made the PDQ it's top priority, even though it was more difficult and costly to produce and had a lower profit margin than Montpelier's traditional line of trucks.

Much like Kay Cox, Nan Mitchell began working at Montpelier just a short time after her high school graduation at the tender age of seventeen. Kay, who was one of her best friends and who had only started working at Montpelier Manufacturing a few weeks earlier, recommended her. Just like Kay, she learned to wear several hats, serving as switchboard operator, receptionist and secretary. Even though she had some idea of what to expect, when confronted with the full force of Mr. Schwartz's unique brand of office etiquette, she admits that he, "Scared me to death. I had no problem with anybody else. In fact, it was really an easy job. He was the type that when he told you to get somebody *(on the phone or out on the shop floor)*, you got them. If he asked me to get somebody on the phone, I did not hook Mr. Schwartz up until that person was on the phone. Mr. Schwartz didn't wait. When Mr. Schwartz picked up that receiver, they had better be there. The person he wanted to speak to could wait but he couldn't. Even if it were a prime customer, it made no difference. *They* would be the one to wait on him; he would wait for *no one*."

When asked why Schwartz had such an attitude, she simply replied, "Because he was Mr. Schwartz." Nan would get nervous if someone Schwartz wanted to see or talk to took their time in responding to a shop floor page or a phone call because she knew that she would soon be the target of his wrath. Those who were most frequently summoned from the shop floor for an "on the carpet" appearance in Schwartz's office were the plant superintendent or one of the foremen and she admits that the summons was most often for nothing more than the routine exchange of information. However, if Nan knew a reprimand was forthcoming, she'd do her best to give the poor soul some sort of advance warning.

There was, however, a good side to Harry Schwartz as evidenced by occasions when he would morph into a very sympathetic and compassionate man. On one occasion when a large government order brought more business than the plant could accommodate, workers were being hired wherever they could be found. One particular man, in sorry shape and in desperate need of a job, asked for an advance on his pay immediately after he was hired, explaining that his family was in dire straits and the money was needed for basic necessities right away. Schwartz gave him the advance without any hesitation before he had even worked his first hour at the plant.

The fact that Harry Schwartz balanced his high expectations and heavy-handed style of leadership with a charitable, almost paternalistic attitude left a few employees awestricken and many employees with a

A circa 1963 Urban Delivery built for the Railway Express Agency. The wide wheel housings indicate that this is a 1½-ton or larger truck. Notable features include a partition with sliding door between the cargo and driver's compartment, translucent FRP top, roll-up rear door, cargo compartment light mounted to the partition, receptacles in the interior sidewalls for retainer bars to minimize cargo shifting and diamond tread plates added to the rear corners of the body for damage protection in close quarters.

healthy respect for him. Nevertheless, Nan characterized the overall office atmosphere as tense. The questions always lingered: "When is he going to want something?" "When is he going to yell at me?" "Have I got my running shoes close at hand?"

Schwartz's office was at the north end of the office building with the switchboard, Nan's primary workstation, right outside his office. When he would clamor, "I want so-and-so!" or "Get me so-and-so!" she began her effort to comply with his order almost before it was fully enunciated. The switchboard was also close to the office entrance. Nan kept a stack of employment applications nearby and handed one to prospective employees as they stepped through the office entrance. A hallway served as a buffer between the office and shop floor. Job applicants would sit in chairs in the hallway as they filled out their applications. Visitors for Harry Schwartz, Bob Treer, Guy Priest, or others in Montpelier's sales department would wait in these chairs until Nan would escort them to the person with whom they came to meet. Nan could view activity in the hallway through a wall of glass windowpanes that started about four feet above the office floor and terminated near the office ceiling.

Although Schwartz's temperament had all but the most callous of the office staff on edge, Nan never felt that his expectations, orders or "requests" were unreasonable. Her problem was with the terse manner in which they were issued.

Nan and a few of the other office workers called Schwartz "Schwartzie" among themselves, as long as Schwartz was out of earshot. In or near his presence, he was always addressed as Mr. Schwartz.

By 1962, Schwartz had retired and Nan's administrative support responsibilities were shifted away from Roy Sjoberg, Schwartz's immediate successor, to Guy Priest, the Montpelier Division's sales manager, and Assistant Sales Manager Lyle Shoup. Sjoberg's stint at the helm of the Montpelier Division lasted little more than a year but it was long enough for Nan and others to characterize him as "another nasty one." Most of her time as the sales department secretary was devoted to preparing price quotations, sending out sales literature and typing letters to customers. Her interaction with customers was minimal. She received work assignments from Rundell, Priest, and Shoup after they had conducted sales activity with customers.

Nan took advantage of the six-week maternity leave offered by the Montpelier Division in April of 1964 to have her second child. By that time, Sjoberg had ensconced himself in the more prestigious confines of White's corporate headquarters in Cleveland and was succeeded in the general manager's position by John Rundell. Rundell was a veteran of seventeen years in Montpelier's sales department and had worked his way up through the ranks of sales manager and sales vice-president. Nan returned to work for a solitary day in June and realized in her heart that she couldn't ship two small children off to a baby-sitter and go back to work, so she called Mr. Rundell the next day and explained her situation. Rundell replied, "Don't worry about it. As of right now you're no longer working for us. If you need a job anytime, give me a call." Nan drew a full six weeks pay while on maternity leave, even though she only returned to work for a single day after her leave expired and wondered if she was entitled to keep the income she received while on leave. Offering somewhat tentatively to return it, Rundell told her to keep the pay she received while on leave and not to worry about it.

Years later, after Nan could compare her work experiences at the Montpelier Division with subsequent office jobs held elsewhere, she could say with all honesty that, "They were good to me. They were very patient with me and kept me on as an employee after my first child and I both had medical problems shortly after he was born. I have no complaints about the way I was treated. Montpelier Manufacturing was a very nice place to work and Mr. Rundell was the best boss anybody could ever have. It wasn't an office atmosphere. Everybody chimed in. If there was work to do, everybody pitched in and did it. It was good working with White, too. I had no problems at all with the guys that came down from Cleveland. In fact when I had my baby, they sent me flowers."

In contrasting the management style of John Rundell with that of Harry Schwartz, Nan conceded that both Schwartz and Rundell were equally effective at the helm of the company in spite of their starkly contrasting management styles. In culinary terms, one resorted to vinegar while the other found sugar a more effective ingredient for motivating people and achieving objectives.

16

A Return to Ashes

In July 1965, the Montpelier Division was working at well over one hundred percent of the plant's normal capacity, finding whatever way it could in a hit-and-miss attempt to meet delivery commitments. Behind schedule toward the end of the month, the plastic molding department was operating with a small contingent of three workers on a third shift in the early morning hours of July 28. Their intent was to mold six fiberglass reinforced tops per night, a feat that could be accomplished by mixing a little extra hardening agent in the polyester resin, working at breakneck speed, and not encountering any problems.

A fiberglass mat had just been laid into a large top mold over a layer of resin and the crew decided to take their nightly ten minute break at about 4:55 AM. The men drew a cup of coffee and had just seated themselves in the break area when they heard a whistling sound, indicating a sudden air leak. Bill Gipple was the most senior man on the small night crew, having spent a full five years working in the plastic molding department. The workers were not afforded a direct view of the plastics department from the employee break area but the noise sounded as if it were coming from that area

Carlton Gray, a temporary hire while home from college for the summer, was at the scene of the fire a few minutes before the Montpelier Fire Department arrived. He jumped aboard the company's aging tow motor to pull a truck chassis out of the fire's rapidly expanding grasp. He then succeeded in towing about a dozen completed trucks to safety.

of the shop so Bill headed back to investigate. He had proceeded less than half-way when it became obvious that a well-established fire had scorched an air hose and caused it to burst and produce the whistling sound. Bill hollered for the other men and continued to approach the work area but soon realized that there was nothing they could do. The fire had already overwhelmed much of the plastics department and soon might begin setting off explosions as containers of volatile chemicals yielded to the fire's intense heat.

In leaving the break area to investigate the cause for the loud whistling sound, Bill explains that, "I never got back there. I saw the light from the fire glow out into the main area of the shop. I probably got about half-way and went back." Bill tried to call the fire department from Superintendent Denver Miller's office but had trouble. "The phones didn't work right or the switchboard might have been shut down." Bill ran outside and yelled to Paul Lower, Montpelier's night watchman, to run over to the Bauer residence on Randolph Street to call the fire department. His efforts to reach the fire department were unsuccessful but employees of the Norfolk and Western Railroad whose tracks were located along the plant's south property line saw the fire and called the fire department at 5:05 AM.

According to Bill Gipple, "An air line had apparently burned in two wherever it started. I headed back that way but it started going fast. I mean, it was really moving fast. Things were blowing up. That fiberglass would explode. I just told the men, 'Let's get out of here!' so we took off and went outside." When asked what fire fighting equipment was available to combat fires within the plant, Bill replied, "I don't think there was anything there. There might have been some equipment back toward different locations in the plant but right there at the fiberglass area, I don't think that there were any on the posts at that time." Unlike the rest of the facility, fires in the plastic department were infrequent; employees were very conscientious, taking the necessary precautions to avoid fires or dousing them as soon as they occurred with water drawn from a sink in the department.

"Looking back, it was a firetrap." Bill added. "If you added too much hardener, the resin would sometimes smolder and char until the resin hardened but it would not ignite." When it smoldered, the resin emitted a rancorous odor. Bill does recall using a bit more of the hardening agent that night to speed up the molding process.

"I don't know how it ever happened so fast. I haven't figured that out yet. I don't think that we were in the break area for more than five minutes because the break, if I remember right, was only ten minutes."

As the first man to observe the fire, Bill was asked many times where the fire initiated. "I don't know where it started. I wouldn't have the slightest idea where it started. You see, it had a wood floor in there, too, but that wood floor was covered with fiberglass where it was spilled on the floor over the years." Nothing appeared to be out of the ordinary as the men worked that night and Bill can think of no unusually hazardous conditions present that could have caused the fire.

Carlton Gray was a college student at the time of the fire and was hired for the second consecutive summer as a temporary worker. After dropping his father, a firefighter, off at the fire station, Carlton drove up Randolph Street to the plant and was one of the first citizens on the scene. He was in possession of the keys to "Nellybelle", the factory's aging towmotor and took it upon himself to pull more than a dozen finished or nearly finished REA trucks to the far reaches of the property as the firefighters began the task of bringing the fire under control.

Henry Pape was awakened shortly after 5:00 AM that morning by a phone call from night watchman Paul Lower and was informed that a fire was raging at the plant. Looking through his upstairs bedroom window, he observed an orange glow in the direction of the plant. Speechless for a few seconds, he then turned to his wife, saying, "She's going this time. I'm out of a job."

Montpelier's Assistant Sales Manager Lyle Shoup recalls, "My father-in-law was one of the local undertakers and whenever the fire department was called to a fire, he would show up to serve coffee and donuts. My phone rang shortly after the fire started. It was my father-in-law informing me that 'White Motors is on fire!'." Lyle's home was just a few blocks away from the plant. A faint smell of smoke provided bleak confirmation of his father-in-law's words. He stepped outside and observed a towering plume of black smoke, backlit by the fire's massive glow. He quickly got dressed and forced himself to head down to the plant, dreading what he would find and wishing in vain that he and all of Montpelier could turn the clock back and get another chance to relive the past few hours, to be given the precious opportunity to extinguish the fire at the very moment that it started.

Salvaging items through the windows on the north side of the Montpelier Division office building became a priority, even as firefighters battled the factory blaze just a few yards away.

Then and now, Lyle describes the fire as, "A roaring uncontrollable fire. Fiberglass, acetone, methyl-ethyl-ketone and other solvents were in there and virtually the whole facility was wood. We had the fire department down there frequently. Many other times, the workers would put out fires themselves. People always said, 'Someday this thing is going to go up in flames'." Explosions were occurring at random as workers assembled and watched firefighters do what they could to contain the fire. One of the acetylene tanks outside under the overhang by the chassis department burst and violently propelled itself into Jack Shaffer's nearby toolbox.

Paul Houk lived on a farm northwest of Montpelier. His aunt and uncle lived on the opposite side of the railroad tracks that ran along the south edge of the Montpelier Division's property. Paul was in the middle of his breakfast when his aunt called and said, "I don't think that you'll have to go to work today, the factory is on fire." Paul went outside and could easily see, in spite of the five mile distance, a dark cloud hanging over Montpelier.

Kay Cox's phone rang as she was preparing breakfast for her children. Her husband picked up the phone and received word from General Manager John Rundell that the body works had burned overnight and that a meeting would be held that morning at Rundell's home. The family looked outside in the direction of the plant and was greeted by a closer view of the black cloud that Paul Houk had just observed.

Al Goebel heard the fire sirens and saw the large black smoke cloud a few minutes before receiving a phone call that confirmed his fears about a fire at the body works. Al was grateful that he was among four shop workers that were asked to stay on and help the division's three shop superintendents finish building or salvaging trucks that hadn't been destroyed. He was disappointed that the fire had consumed roughly $600 worth of completed work tickets, the equivalent of six weeks of wages, that he reluctantly had to bank over the previous few months of hectic production activity.

The office staff and department supervisors assembled at Rundell's home that morning, many of whom just sat and cried, as did John Rundell himself. Preliminary plans were made to notify White corporate headquarters and each of the division's customers. A few days after the fire, the office and engineering staff went through files. Fortunately most of the files and engineering drawings were salvageable. The office building, built in 1943 and modestly enlarged after the White acquisition, was of glazed brick and concrete construction and fared much better than the wooden factory building.

An impromtu relay team retrieves engineering files from the basement of the office building through a rarely used door that affords basement access. Additional files would be salvaged later by wading through fifteen inches of water that inundated the divison's engineering office.

As soon as it was safe to enter the office building, fire fighter Bill Schatzer gave Montpelier Division draftsman Clare Dilworth a pair of fireman's boots that were too large for any of Montpelier's firemen. Clare used them to wade through fifteen inches of water in the engineering department, located in the basement of the office building. He salvaged whatever engineering documents he could. With some additional help the blueprint machine was extracted and temporarily moved to Keith Miller's auto body repair shop. Clare still has those boots and whenever he finds an occasion to use them, he thinks back to the circumstances under which they were acquired.

Keith Miller offered his newly acquired body shop building as a temporary location for performing repair work on damaged trucks and completing the final phases of work on nearly finished trucks that had been rescued from the fire, but there wasn't actually much work that could be done at his facility. Most of the truck salvage and finishing work was done on the factory premises in a couple of storage sheds that were pressed into service as primitive manufacturing sites after electrical power was made available on a temporary basis. The weather was largely cooperative, allowing workers to set a few truck bodies outdoors on barrels in order to work on them.

The north wall of the Montpelier Division manufacturing plant is enveloped by the intense flames. The fate of the Air Force ambulance perilously close to the wall is uncertain at this time.

A large truck order for the Railway Express Agency was being processed when the fire took its toll. The few workers that were retained focused on fulfilling what they could of the company's commitment to REA. Montpelier High School served as a temporary office facility for a downsized office staff. The Aro Corporation of nearby Bryan, Ohio donated used office furniture to the beleaguered office workers.

Writing on the Wall

Lyle Shoup admits, "We had the fire department down there frequently and there were many fires that workers would put out themselves." As superintendent, Henry Pape enlisted the aid of two or three employees who were volunteer firemen and organized a fire brigade. The fire brigade and the procedures they instituted actually saved the plant from serious damage on more than one occasion.

There was no sprinkler system in the plant. Unfortunately, the urgings of Margaret Schwartz (the lowest ranking and most docile member of the company's board of directors in the '40's and '50's) to erect a water tower and install a sprinkler system had gone unheeded by her husband, even though the city had extended the Randolph Street water main up to the factory property in 1937 for that very purpose. Upon learning of the loss, Harry Schwartz was quoted by his son Jerry as saying, "There's my life, in ashes." Harry was happily retired, remarried and living in far away Phoenix, Arizona at the time of the fire but was still crushed by the event.

Denver Miller recalls that a downhill slope behind the factory was a dumping ground for rejected FRP panels, excess resin, resin that was unfit for use and resin that had accumulated on the floor of the plastic department. Workers in the plastic department would dump the resin and hardener mixture on the ground and the resin, grass or adjacent materials would occasionally ignite. The Montpelier fire department grew tired of the plant's frequent calls and would have preferred to have left the annoying firm to its own devices, if duty and obligation hadn't overruled. The city wouldn't allow the company to borrow the fire department's hoses but a body works employee had contacts in the fire department of the neighboring town of Edon

Concrete posts that once supported roof joists stand as the sole monuments to thirty-eight years of industry.

and obtained a fire hose that would enable the employees to douse fires that erupted due to the discarded resin. Many calls to the local fire department were thus avoided and the company's reputation was not quite as besmirched.

Yet there were times when the fires that erupted were more than the fire brigade could handle. The May 14, 1964 *Leader-Enterprise* reported that "Montpelier firemen poured about 100,000 gallons of water on the White Motor Company dump on Tuesday and Wednesday." The January 28, 1965 edition reported that "Alert action by a fellow employee saved Claren Keith from incurring still more serious burns after an explosion at the Montpelier Division of White Motor Company on Tuesday. According to reports, some paint thinner exploded, igniting Mr. Keith's pants. Virgil Bashor, who was nearby, took off his shirt and wrapped it around Mr. Keith's legs, putting out the blaze." Keith suffered first and second degree burns and was treated and released from the hospital the following day.

On July 12, a mere sixteen days before the fateful fire that destroyed everything, the *Leader-Enterprise* reported that "A tractor and shed at the White Motors plant caught fire this morning while a number of barrels of paint thinner threatened to explode at any minute. Montpelier firemen managed to bring the flames under control, causing a minimum amount of damage. Albert Baden was on the tractor at the time and jumped to safety when the fire started."

The large green and white metal sign that proclaimed these premises as the "White Motor Company, PDQ Division, Montpelier Body Division" collapsed into the roof of the office building moments before this photo was taken, sending a subliminal message that the sign's affirmations were no longer true.

Aftermath

Naturally, most of the employees were dismissed immediately. Kay Cox stayed on for a month, making many phone calls, organizing files, and helping John Rundell and Buzz Bartick assemble what they hoped would be a very persuasive reorganization plan to present to White Motor Company. Guy Priest and Lyle Shoup, the core of the Montpelier Division's sales staff, stayed on in anticipation of a viable reorganization plan. Paul Strawser, Henry Pape and Denver Miller led the truck salvage effort on the factory lot, doing their share of work along side four hand-picked men.

Within a week of the fire, many of the Division's former employees filed for unemployment compensation. Eligible employees received $42 a week, plus $5 for a dependent wife and $3 each for up to two dependent children, the maximum amount being $53 weekly.

Putting the loss in monetary terms, the factory was valued somewhere between $1.5 and $2 million. The 250 employees of the plant had a payroll that exceeded $1 million annually and the tax valuation of the facility was about one-half million dollars.

A thought that crossed the minds of a steadily increasing number of Montpelierites was that the fire was set intentionally. Presumptive blame could easily be placed on White Motor Company because of their inability to turn a profit and sell significant numbers of Montpelier-built White PDQ trucks. White added fuel to these suspicions in the first few weeks following the fire by withholding assurances that the factory would be rebuilt. No facts, however, even came close to incriminating White Motor Company or anyone else that the fire was deliberately started.

The metal shed was destroyed more from exploding containers of volatile chemicals than from the fire. Stacks of bumpers were salvageable items but whether they would find their rightful place on the Montpelier-bodied delivery trucks remains to be seen.

17

Montpelier Struggles to Recover

Apart from the feverish pace during World War II Montpelier Manufacturing had never been busier than it was at the time of the devastating July 28 fire. With a significant order backlog, prospects for continued business were extremely bright. It appeared that the popularity and profit potential of the modern new Urbans would more than offset the burdensome cost of producing and attempting to sell White-branded PDQs. Montpelier was accustomed to filling smaller orders for predominantly local delivery truck fleets, but the company had succeeded in acquiring large orders from dominant companies like Continental Baking, Nickles Bakery, Jewel Tea Company, the Railway Express Agency, and United Parcel Service - companies that utilized thousands of delivery trucks.

The brief but phenomenal success of the Montpelier Division's newest aluminum and steel bodied forward control delivery bodies, developed jointly with Continental Baking and other major customers, compelled an envious White Motor Company to demand a version of its own. The result was the White Aluminum-PDQ, released in the Spring of 1965, only a few months before the Division was destroyed by fire. Customers could choose between Chrysler-built Slant Six or 318 CID V-8 gasoline engines or Perkins 4-236 and 6-354 diesels.

C. A. Lehman & Sons, Inc.

901 HAYDEN STREET
FORT WAYNE, IND. 46803
TELEPHONE 219-742-1461

September 14, 1965

White Motor Corporation
Montpelier, Ohio

Attention: Mr. Glen Taylor

Dear Sir:

We offer the following proposal for your consideration. We will Furnish all labor, material, plans, engineering and state approval necessary to complete the new manufacturing plant for the sum of Four hundred and two thousand dollar ($402,000.00). $4.27 sq. ft.

We include the following basic specifications:

1. Manufacturing area to be 180' x 500' with 18' ceiling.
2. Office area to be 50' x 85' with 10' ceiling.
3. Sidewalls to be concrete blocks 4' high with insulated metal panels above and translucent panels as needed.
4. Roof to be insulated tar and gravel, 20-year bonded.
5. Floor-5" concrete, 3000 psi with 6 x 6 #6/6 reinforcing.
6. Overhead doors to be wood with one section glass and equipped with electric operators.
7. Passage doors to be metal 3' -0" x 7' -0".
8. Electric lighting to be industrial type fluorescent fixtures to maintain 40 foot candles at normal working height.
9. Employees wash room to have 10 stools, four urinals, and five Bradley wash basins. Office washrooms to have stool, urinal and wash basin. Rest rooms to have metal toilet partitions.

This letter was one of three preliminary cost estimates the Montpelier Division received from contractors for a new manufacturing facility and office building. Handwritten notations indicate that Glen Taylor contacted C. A. Lehman & Sons to determine earliest possible dates for the start of construction and the project's anticipated completion. Itemized details in the Lehman letter describe a facility that would enable the Montpelier Division to shed the burdensome handicaps associated with its outdated plant and bring them ever so deservedly into the modern age of manufacturing.

In spite of Montpelier Division's bright future, White put forth little effort to save it after the fire. Efforts to revive the Division were left to the workers in Montpelier. Draftsmen and engineers had little work to do for customer's truck orders, so Glen Taylor set them up in the front of Keith Miller's body shop and led an effort to design a new plant. By August 26, 1965, the group completed an architectural plan view of a new 500 foot x 180 foot factory floor and adjoining 85 foot x 50 foot office building. The plan provided generous space for metal forming machinery and a shop floor with enough space for six chassis in the chassis prep department, twenty-one bodies in various stages of construction in the welding department, a centrally located priming and paint department and capacity for eighteen bodies in the final assembly and body drop area.

Mr. Glen Taylor   Page 2   September 14, 1965

10. Heating to be with unit heaters gas fired and to have Capacity to maintain 70° at minus 10°.
11. Plumbing to include auxiliary water lines, drains for air, compressors, paint facilities and fire protection.
12. General office area to be paneled, acoustical ceilings, aluminum entrance door, vinyl asbestos tile, 90 foot candles lighting, forced air heat and air conditioning, and windows covering 30% of the exterior walls.
13. Exterior of building to be painted and sealed. All exposed wood in office area to be painted three coats.
14. Exterior wainscoting of office to be face brick.
15. Loading dock and ramp to be as shown.

This is intended to be an outline specification only. We have not gone into great detail to itemize each item of construction. If we are the successful contractor on this project we will furnish complete plans and specifications for your approval prior to the starting of construction.

We thank you for the opportunity of quoting this work and hope that we may be of further service to you.

Very truly yours,

Dick Lehman, President
C. A. LEHMAN & SONS, INC.

DL:lz

Start Oct. 1/65
Inside by 1st Jan 1966 in Factory
Complete Office — 15th Feb. 1966

MONTPELIER MANUFACTURING

A DIVISION OF THE WHITE MOTOR COMPANY
MONTPELIER, OHIO 43543

J. C. RUNDELL
GENERAL MANAGER

September 28, 1965

Mr. Charles F. Meyers
President
Hercules Galion Products, Inc.
Galion, Ohio

Dear Mr. Meyers:

Listed below are the names and addresses of the companies with which we have done recent body business at Montpelier.

Chicago Daily News	404 N. Wabash, Chicago, Ill.	~~Mr. Ed Finn~~
Chicago Sun Times	311 S. Clinton St., Chicago, Ill.	Mr. Cliff Sullivan
Chrysler Corporation	P. O. Box 2088, Detroit 31, Mich.	Mr. W. R. Ball
Continental Baking Co.	P. O. Box 731, Rye, New York	Mr. Walter Chanda
Ford Motor Company	P. O. Box 658, Dearborn, Mich.	Mr. Al Weddle
Railway Express Agency	219 E. 42nd St., New York 17, N. Y.	Mr. Frank Cousar
Illinois Bell Telephone Co.	309 W. Washington St., Room 100, Chicago 6, Illinois	Mr. John Boyd
Marshall Field & Co.	601 W. Polk St., Chicago 7, Ill.	Mr. E. R. Behnke
Alfred Nickles Baking Inc.	26 E. Main St., Navarre, Ohio	Mr. Maurice J. Eckroad
The Philadelphia Inquirer	400 N. Broad St., Philadelphia, Pa.	Mr. J. H. Houston, Jr.
Rubel Baking Company	3025 Bathgate St., Cincinnati, Ohio	Mr. Robert Thomas
United Parcel Service	643 W. 43rd St., New York, N. Y.	Mr. Ledyard Pfund

In the listings we have made above, we have included the names of the Fleet Superintendents or Purchasing Agent who is responsible for the various accounts in case you would care to call any of these people.

Sincerely,

J. C. Rundell

JCR:jkc

TELEPHONE (419) 485-3161 TWX 419 - 922 - 0702

By September 1, the group produced a "Proposed Plot Layout," a plan view of the new factory on the existing Randolph Street property using existing outbuildings and concrete pads wherever feasible. A "Proposed New Plant Plot Layout" for a new "greenfield" factory at the intersection of East Avenue and East Brown Road, about one-half mile southeast of the company's existing Randolph Street property, was completed on September 22. A "Machinery & Equipment Layout," a plan view showing the arrangement of welders and metal forming equipment in greater detail, was also completed on September 22.

Blueprints of these drawings were forwarded to building contractors as soon as they rolled off the blueprint machine that was salvaged from the July 28 fire. Building contractors must have been looking over the shoulders of the Montpelier Division draftsmen as they drew the plans because no less than three well-detailed preliminary cost estimates for new construction were received by September 14. The September 14 written estimate received from C. A. Lehman & Sons, Fort Wayne, Indiana, has been reproduced and appears on previous pages. Penned notations on the C. A. Lehman & Sons estimate, apparently jotted down by Glen Taylor in a follow-up phone call, indicated that Lehman & Sons would be available to break ground by October 1, have the factory portion of the new facility available for occupancy by January 1, and complete the office quarters by February 15. In less than six months and for the sum of $402,000 ($4.27 per square foot) the Montpelier Division could settle into a modern concrete and steel building that would finally bring them completely out of the days of farm wagon production.

The Division's haste in drawing up new plans and obtaining construction estimates had a three-fold purpose. First and foremost, it was their duty to demonstrate to their higher-ups at White headquarters in Cleveland that they weren't capitulating to the devastation wrought by the fire. They were hoping that the tenacity, resiliency, resourcefulness, and speed with which they assembled a recovery plan would win White Motor Company's endorsement and instill a willingness to apply the insurance proceeds and whatever additional sums of money were needed to start afresh. Failing that, a core group of Montpelier employees were eager and willing to declare their emancipation from White Motor Company. Hoping to drop the burdensome purchasing, manufacturing, and logistics headaches associated with the low profit PDQ trucks into White's corporate lap, they envisioned an employee-owned company free of the confining dictates of its overbearing parent. Finally, those who were leading the recovery plan wanted to put together a viable proposal quickly in order to deter many valuable employees from seeking employment elsewhere. A buyout had been on the minds of those employees who realized White's ownership of the company had been exploitive and that the division's future was arguably less secure than it had been before White bought the business

Interest in resurrecting the business in the Montpelier community was understandably strong. The pool of talented employees was still reasonably intact and the company had a strong and loyal customer base headed by UPS, Continental Baking and the Railway Express Agency. The Montpelier Chamber of Commerce implored White Motor Company to rebuild, but the chamber itself had no financial resources to help with reconstruction. Another (somewhat ill-conceived and unfortunate) reason that the chamber of commerce and local government abstained from efforts to help the downtrodden company was a fear that any show of favoritism would alienate other local employers. A few prominent businessmen from Edgerton, a village located about fifteen miles southwest of Montpelier, talked informally with some of the employees, encouraging them to relocate to a vacant factory building that was available in their community.

After receiving the draft copy of the White Motor Company news release that the Montpelier Division would not be rebuilt, John Rundell wrote this letter to Charles Meyers, President of Hercules Galion Products. Hercules Galion had contemplated buying the remaining assets of the Montpelier Division but decided against it a few day earlier. As a simple courtesy, Rundell provided them with this list of Montpelier Division customer contacts, knowing the Montpelier Division would have no further use for it.

WHITE EXECUTIVE TRUCKS

NEWS BULLETIN

Advance copy of news released to the press by:
WHITE TRUCKS division of The White Motor Corporation • Cleveland, Ohio 44101

FOR IMMEDIATE RELEASE

FROM: DONALD B. SEEM
WHITE TRUCKS, A DIVISION OF
WHITE MOTOR CORPORATION
842 East 79th Street
Cleveland, Ohio 44101
431-2000

CLEVELAND, OHIO - October 4, 1965 ... Announcement that White Motor Corporation will not rebuild its Montpelier, Ohio manufacturing facilities destroyed by fire on July 28, was made today by H. J. Nave, President of the White Truck Division.

Nave, in a letter to all customers with PDQ trucks on order wrote, "I take this opportunity to thank you for the business you have given us and assure you that parts for PDQ's will be available for a reasonable length of time through your regular White Trucks' Service Centers."

In a similar letter to customers with orders for Montpelier bodies, Nave wrote, "We regret that we will not be able to fill your present order and that we are not able to continue manufacture of these bodies in the future."

#40091A

White Motor Company eventually released the news of their intentions not to rebuild the Montpelier Division to the media on October 4, 1965, six days after John Rundell received a draft copy of the announcement. The amount of the insurance settlement that White received wasn't disclosed.

Montpelier Manufacturing's office building survives today, standing surprisingly intact after forty years of vacancy and neglect. The former office entrance is in the dark recess in the center of this photo. A molded concrete header over the doorway insistently declares the building as an "Office" in bold block letters. A crude sheet metal addition to the building was appended toward the west at some time and a business that struggled to produce cellulose insulation from old newspapers operated in the structure for awhile. A wilted stack of newspapers and nature's elements are the only items housed in the facility today. The property owner erected a chain link fence around the building in the late 1990's to keep himself and others out of trouble.

Handwritten notations dated September 3 by Buzz Bartick, the Montpelier Division's Director of Planning, provided data on the total number of trucks shipped in 1964 and in the seven months leading up to the fire in 1965. Also included was the number of units that the company anticipated building throughout the remainder of the year. No one knows at this time what logic Bartick used as a basis for anticipated truck volumes for the remainder of 1965. All of the numbers were likely used to convince White Motor Company and other possible stakeholders that business had been very brisk leading up to the fire and to advocate rebuilding the plant to capitalize on the division's large order backlog and the company's long-standing reputation among its customers.

One factor that may have dissuaded White from initiating a rebuilding plan was the discovery by White's auditors that some rather significant accounting errors were found in the Montpelier Division's financial records. While the nature of these errors were never revealed in detail, it's possible that White discovered that PDQs were still being sold at a significant loss, contrary to the Montpelier Division's monthly financial reports which indicated that they were marginally profitable. The monthly reports reveal a positive profit margin for PDQs, even though the majority of knowledgeable Montpelier Division employees understood that, in Denver Miller's words, "The PDQ never made a dollar's profit." Upon finding the accounting errors, White Motor Company promptly dismissed the Montpelier Division comptroller.

Employees were generally optimistic that White would give them the authority to proceed with a recovery plan in the first weeks that followed the fire. Insurance investigators met with Bill Gipple, the ranking employee on duty when the fire broke out. Bill told them everything he knew, which wasn't an awful lot. He pointed out once again that, "When welding torches and other hazardous items started to explode, the fire department was hampered from fighting the blaze." The number of trucks and chassis in the area surrounding the plant further hindered access to the fire. The fact that insurance adjusters were on the job bolstered hopes that a claim settlement would be forthcoming and that it would provide a substantial down payment on a new facility. White Motor Company was besieged with letters from many of the Montpelier Division's customers and suppliers as well as local and state officials stressing how important a rebuilding program was, not just to Montpelier's economic health, but to all who did business with the company. A representative from the state Industrial Development Department joined several Montpelier leaders in visiting White officials in Cleveland during the week of August 9 to discuss rebuilding plans.

Each passing day of White's reticence further eroded the hopes and ambitions for the turnaround envisioned by Rundell, Taylor, Bartick, and the employees still on the payroll. Finally, on September 28, exactly two months after the fire, Rundell received a draft copy of a White press release indicating that White Motor Company was not going to rebuild their Montpelier Division, electing rather to simply bank the check from the insurance settlement. That same day a dejected Rundell, signaling his personal surrender, sent a letter to the president of the Hercules Galion Products Company, providing his former competitor with names, addresses, and personal contacts for each of the Montpelier Division's prime customers, apparently fulfilling a recent request for the information. A copy of that letter, perhaps the last one written on company stationery, can be found on a previous page. Rundell knew that Assistant Sales Manager Lyle Shoup had a standing job offer from the Aro Corporation in Bryan, Ohio and advised him that day to accept it. He informed the *Leader-Enterprise* of White's decision and a bold front-page announcement of the unfortunate news appeared in the September 30 edition of the newspaper, four days ahead of White's eventual news release.

Editor Bill Bryce's column began with an account of John Rundell's visit a day earlier and his report that "This is the end of the end." Bryce went on to remind readers:

"Under Mr. Rundell the company increased sales and production tremendously, and that Montpelier took pride in this industry and its more than two hundred employees.

"When the fire last July completely destroyed the plant, the community optimistically predicted an early rebuilding. But, as White held off making a decision to erect a new factory, gloom settled in.

"The day of the fire, the Montpelier Area Chamber of Commerce, the Montpelier Development Company and a number of local businesses and individuals phoned or wired White officials in Cleveland, urging the company to rebuild and offered assistance.

"Trips were made, costs were obtained, proposals were submitted - and still White declined to make any commitment.

"Then it became known that a firm in Galion was interested in purchasing local assets of White - being especially interested in the labor here that is so skilled in this type of production. Hope was on the rise again and the Development Company and the Chamber worked hard to arrange a program of construction and financing that would satisfy prospective new owners.

"Yesterday it was learned that this prospect had turned down the proposition and that White had decided to discontinue the operations completely."

In all, it had been an extremely rough year for John Rundell. He had suffered a serious heart attack in September 1964 and another in December, just as his condition began improving measurably. During his convalescence, management of the Montpelier Division was left in the capable hands of forty-one year old Assistant General Manager Bill Sayers but, as noted earlier, Bill succumbed to a fatal heart attack in the presence of his fellow employees in May 1965.

Dejected but not taken by surprise, salaried personnel received varying amounts of severance pay. Lyle Shoup recalls that his severance pay amounted to about $350. Denver Miller, Henry Pape, Buzz Bartick, and Bob Treer were invited by White Motor Company to come to Cleveland for job interviews. Bartick and Treer accepted job offers and began work at White facilities in Ohio and Pennsylvania. Denver Miller declined, preferring to find work locally to keep his children in a familiar school setting and continue farming on the side. Henry Pape decided that he was not interested in working in Cleveland after one evening of battling Cleveland's rush hour traffic. Chief Engineer Glen Taylor accepted an engineering position with the J.B.E. Olson Company, a Montpelier Manufacturing competitor in Sturgis, Michigan.

Nothing has been built on the precise location where the original Montpelier Manufacturing factory once stood. The concrete foundation and footings that once supported the wooden structure that started out as a factory to produce Waggoner Talking Machines still exists. A fledgling business that processed old newspapers into cellulose fiber insulation occupied the glazed block office for a while and a ramshackle sheet metal addition was appended to the office building's west side. Neither the business environment nor the building were hospitable to the enterprise and it was short lived. Looking through the open doorway through which passed Harry Schwartz, John Rundell and the Montpelier Manufacturing office staff, a small pile of old damp newspapers is still visible, residue left from the last day the office building served anything approaching a useful purpose.

The employees of Montpelier Manufacturing proved themselves to be an indomitable and resilient group of people. Those who departed, whether after the fire or in the years leading up to it, had the necessary skills and talents to prosper in other vocations. Some sold shoes, insurance, or investment securities. Others found good jobs with the state highway department or other industries in the area. Almost without exception, those who survive today are in excellent spirits, live in comfortable homes and were exceptionally kind, cooperative, and friendly to a stranger who decided that the story of their labors and the labors of their loved ones needed to come to light. The following letter, written by the wife of a long time Montpelier Manufacturing employee who lost his job on the day of that fateful fire, appeared in the October 14, 1965 edition of the *Leader-Enterprise*. Reading it will convince even the most hardened skeptic that *there truly is beauty in ashes*.

To: Editor, Leader-Enterprise

Dear Sir:

Other obligations have kept me busy, but I feel it's time I "voice" a word of thanks to our village leaders.

As the wife of a White Motor Company employee I would like to thank John Rundell, Gene Kimmel, and all others (many that I don't know) for their untiring efforts in brain work, time from their businesses, trips, etc., all suggestive to the cause of rebuilding in Montpelier.

White Motor Company has commended this group for the assistance they gave, but I feel we, their fellowmen, should also tap their shoulder and say thanks for a job well done - even when the answer was not what we hoped for.

Roscoe Horton, my Sunday School teacher for many years, taught me a verse of scripture adopted by our class as a motto, Romans 8:28, "And we know that all things work together for good to them that love God, to them that are the called according to His purpose." With this thought in mind, maybe the Lord has work for some of us in another location. If this be so, I shall have many memories next to my heart of more than 26 years in Montpelier. Coming here as a bride I feel it has been good to us and our eight children. Love and prayers to all, especially our village fathers.

Glema Treer

Brief Biographies of Montpelier Manufacturing's Most Venerable Employees

Harry A. Schwartz

Harry Schwartz was the engine that propelled Montpelier Manufacturing through the Great Depression and through all of the product innovations that made Montpelier a leading manufacturer of delivery trucks. A history of Montpelier Manufacturing is, in essence, the life story of Harry Schwartz.

Harry Albert Schwartz was born in Montreal, Quebec on February 10, 1896 to his German immigrant parents, Philip and Hazel Schwartz. Not long after Harry's birth, the family settled in Cleveland where Philip and Hazel made their home until 1942. Philip Schwartz became a naturalized citizen of the United States in 1906 and valued his U.S. citizenship so much that he carried his citizenship papers with him at all times.

Harry attended a technical high school in Cleveland with the intent of preparing for entrance into a technical college. He entered Case Institute of Technology in Cleveland immediately after graduating from high school and, based on his lifelong engineering achievements, many have assumed that he was a Case graduate. Actually, Harry didn't pursue a formal degree beyond his freshman year at Case, sensing that he could educate himself in the engineering disciplines much more quickly and efficiently than could the structured program that Case offered. Occasionally, he regretted not having an engineering diploma and would sometimes appear to be a tad jealous of his wife Margaret who held a degree from Oberlin College. Margaret would assuage his frustrations by assuring him that he was far smarter than any professor at Case *or* Oberlin.

Harry worked briefly in Cleveland after leaving Case. Around the age of 20 he moved to Defiance, Ohio to work at the Defiance Machine Works. He soon became deeply involved in the design of machinery used to manufacture World War I anti-aircraft guns. Though Harry had become a U.S. citizen in 1906 he had difficulty obtaining a top-secret government clearance for the war-related work at Defiance Machine Works due to the circuitous route his parents took to U. S. citizenship. But when Harry boarded a bus with other draftees to leave Defiance for boot camp he was ordered to disembark because of his unique ability to contribute to the war effort at Defiance Machine Works.

In the 1920s Harry developed and patented a permanent-mold casting process. In the years that followed, he traveled extensively, securing foreign patents for the process in virtually every developed country of the world. He set up a company to license the permanent mold process and collect licensing fees but never divulged what, if any, income was derived from the patents or any licensing agreements.

Tax returns show that in 1923, 1924, and 1925 Schwartz made roughly $4,000 - quite a substantial sum for a young man in those days. Feeling that the federal government was fully entitled to share in his good fortune, the only business deduction that Harry claimed over that three-year period was a $10.50 train fare from Bryan, Ohio to Cleveland.

Initially, Harry was nothing more than an investor when his ambitious friend, James Robertson established Montpelier Manufacturing in the old Turnbull Wagon Company building in Montpelier. It soon became apparent that Robertson needed more than just Harry's financial backing, so Harry left the Defiance Machine Works and joined forces with Robertson in late 1927.

When asked years later why he left the success and prosperity that came with his position at Defiance Machine Works, Harry facetiously responded, "To protect my investment." Harry's older son James surmises that an ulterior motive for joining Robertson and ultimately taking full command at Montpelier Manufacturing was to be, once and for all, in business for himself.

After working through 1928 without meaningful compensation, Robertson resigned as a company officer and left in early 1929. Establishing a new manufacturing business is difficult in any environment, but the Great Depression made the task all the more challenging for Montpelier Manufacturing. It's a testament to Schwartz's tenacity, engineering know-how, and business acumen that the company survived. Schwartz was able to persuade two prominent Montpelier businessmen, J. D. Hill and G. Grant Stahl, to invest heavily in the business throughout the 1930's in order to offset the company's operating deficits during the Depression.

The business relationship between Schwartz and J. D. Hill soon developed into an increasingly deep mutual appreciation for each other's outstanding qualities. As Harry became acquainted with J. D.'s wife, Grace, and his two daughters, Margaret and Henrietta, a magnetic attraction developed between Harry and the entire Hill family. It suddenly seemed to the five of them that Harry was the perfect complement to the Hill family and that without him, the family would be incomplete. A natural consequence of mutual affections came on August 17, 1929 with the engagement of Harry Schwartz and Margaret Hill.

An October 26 wedding ensued. Stock market death rattles on Wall Street had been alarmingly frequent in the weeks just prior to the wedding and the fateful and devastating market crash occurred a mere three days after Harry and Margaret took their vows. The big day for Harry, Margaret and much of Montpelier escaped the cruel economic realities that were soon to confront the world and the event was a delight for all who attended.

The October 31 edition of Montpelier's *Leader-Enterprise* reported in an artistic manner rarely seen in today's print media that "The First Presbyterian Church of Montpelier was a scene of unusual charm and beauty on Saturday, October twenty-sixth, when Miss Margaret Louise Hill, daughter of Mr. And Mrs. James DeLos Hill, was united in marriage with Mr. Harry Albert Schwartz of Montpelier."

It appeared that Harry was just as intent on becoming Mr. and Mrs. Hill's son-in-law as he was in becoming Margaret Hill's husband. When recently asked about various family ties, Jim Schwartz recalls his father stating that, "If it weren't for your Grandfather and Grandmother Hill, I would have never married your mother." Mr. and Mrs. Hill were highly respected, almost adored by Harry. Harry would do virtually anything that the Hills asked of him and would often say of J. D. Hill, "He's the only man that I've ever met that I can't praise enough."

Harry started out as the company president but relinquished that title to J. D. Hill, primarily because Hill was largely responsible for financing the firm. Harry was relegated to secretary and general manager in name only. In truth, he ruled over all facets of the company, functioning as the firm's chief operating officer. Realizing that the company was highly leveraged, in tough economic times, and still without a well-established customer base, Harry was frugal and ran the company in a conservative manner. He relied on J. D. Hill and Grant Stahl only for longer-range strategic plans. His expertise was engineering. He disdained the sales and marketing aspects of the business but his engineering background was an indispensable asset when it came to closing a sale. Harry's strong engineering background enabled him to determine the feasibility of virtually any customer request "on the spot" and agree to provide it without getting involved in detailed feasibility studies. Harry was the type to bring a preliminary design sketch back to the plant on the backside of a paper napkin, passing it along to a draftsman or the pattern department for initial prototypes.

A sales department gradually evolved after World War II but Harry Schwartz was still the person who made significant sales and customer contacts at the highest levels. The sales department was most often involved with repeat sales, providing after-the-sale support, preparing highly detailed and lengthy bids in response to customer requests for price quotations and contacting prospective customers that had previously requested sales literature by phone or mail.

Whether it was sales, engineering, manufacturing or the company's bottom line, Jim Schwartz recalls that "Nothing happened down there without Harry Schwartz knowing about it." Jim surmises that his father's domineering approach to people and the business may have been one of the reasons that the business wasn't more successful. "He didn't develop a middle management team. He *was* management, period." Harry personally signed all of the payroll checks. Every morning, Schwartz would go down to the post office and personally open all of the company's mail.

As children, Jim, Jerry and Grace Margaret Schwartz were strictly forbidden to put their feet on the furniture at their home. Jim's first summer job at Montpelier Manufacturing was in the engineering department, located in the basement of the office building. The very first time he went upstairs to his father's office to check on an engineering detail, he found his father sitting in his leather office chair with his feet propped up on the desktop. Jim knew better than to ask for an explanation. For Harry, office etiquette and home etiquette were two entirely different and unrelated matters.

Plant Superintendent Bill Sprankell once observed Harry on the phone conversing with an important customer contact. Seated in a wooden chair, Harry was leaning back, once again with his feet on the desktop. The chair suddenly fell apart, sending Harry's feet in the air and his posterior to an unforgiving tile floor with considerable impact. Unfazed by the chair's sudden disintegration, Harry never missed a beat, picking himself up off the floor and continuing the conversation just as if nothing had happened.

Bob Treer was Montpelier Manufacturing's purchasing agent. Harry constantly found fault with Treer and frequently hurled invectives in his direction without the slightest hesitation. Harry found it easy to do so because Bob was a very docile person who never got contentious or argumentative. On one occasion in particular, Harry asked Bob a question and Bob answered, "Yes, but..." and started an explanation. Harry retorted, "Not 'Yes, but', I want a 'Yes' or a 'No'! What are you, a man or a mouse?" Bob replied, "Yes, but..." and tried once again to explain the situation to Schwartz who promptly interrupted him again. The cycle repeated itself one or two more times before Schwartz stormed away in disgust. Treer was relatively small in stature and his mild manner belied the fact that he had an incredible, almost bone crushing grip that much larger men learned to respect after mistakenly thinking that they could subdue it. Jim Schwartz thinks that if Treer would have snapped back at Harry, the verbal attacks would have lessened or stopped completely.

There were times when Schwartz would treat others in an outwardly hostile manner and then find himself getting a dose of his own medicine. On those occasions, Harry would suddenly turn more diplomatic, de-escalate the war of words and extend a reasonable level of respect and courtesy to the person. Harry even capitulated to his son Jim on one occasion when Jim responded with a quick retort to one of Harry's caustic remarks. Retorts were ineffective, however, when there was no doubt that a worker was guilty of a blatant error or mistake. Under those circumstances a worker took his verbal beating, asked for forgiveness and agreed to amend his ways, actions which almost always resulted in a pardon from "Judge Schwartz".

Jim admits, "Dad used to just cripple people with his blow-ups." Harry would dismiss the point of controversy in a few minutes and forget about it but he left people crushed in the aftermath of his tirades. He was the same way to a somewhat lesser extent with his children, a forceful man to say the least.

Jim recalls that the women who worked in the office were discouraged from venturing out onto the shop floor because of all the jeering and whistling. When he worked in the engineering office, young Jim Schwartz was drafted by his female office workers to go out into the shop to purchase snacks and beverages from the vending machines that the office staff had requested. Upon returning to the office, Jim would be confronted with a glaring stare from his father if Harry happened to see him performing the women's work.

Office typewriters would be pecking away at an intense pace while Harry was within earshot. Office workers heaved a collective sigh of relief and the typewriter cacophony dropped considerably when Harry ventured out into the shop. With the first indication of Schwartz's return, the typewriter intonations resumed at their previous frantic pace. If Harry felt that office chatter was excessive or inappropriately lighthearted, he'd rise from his desk and cast a stern gaze in the direction of the likely source of the problem through the glass partition that separated his office from the office pool.

Montpelier Manufacturing's board of directors elected Grace Hill, J. D. Hill's wife, as president upon J. D.'s death in 1939. Both Harry and Mrs. Hill would sign paychecks and checks written to suppliers every day or so. Harry would stop by the Hill residence on East Main Street and discuss business details as she signed each of the checks. In 1953 Harry reluctantly revealed to Mrs. Hill that he regretted not being able to say that he was the company's president and chief executive officer. An astonished and sympathetic Mrs. Hill responded, "Harry, I never gave it a thought!" and the next day Harry became President and Mrs. Hill became Chairman of the Board. Harry wasn't out for titles. He was perfectly content functioning in the capacity of secretary and general manager but he felt that he could be far more effective at networking with executives in other companies with which he desired to conduct new or increased business if he were president. Titles aside, Mrs. Hill was paid a relatively small salary commensurate with her largely passive role in the operation of the business.

In the summer of 1956, Jim Schwartz, now at the age of twenty, was "promoted" from the engineering office to the shop floor where he used rags and solvents to wash oils and die lubricants off of assembled bodies prior to painting. After a full day of inhaling the volatile organic compounds from the evaporating solvents, Jim would come home filthy and staggering drunk from the fumes. A distraught and appalled Margaret Schwartz would ask Harry, "Why does 'Jimmy' have to have the dirtiest job in the plant? Can't you find something else for him to do? Look what it's doing to him!" Harry dismissed her entreaties by asserting that Jim was somehow profiting in many ways from his summer job and that, overall, "It's good for him."

In spite of all the detractions, Jim enjoyed working each summer at Montpelier Manufacturing. When his father sold the business, Jim was somewhat remorseful, feeling that the business had, to some extent, been sold out from under his brother and him before they were experienced enough to play a significant role in the company's future.

Jim's younger brother Jerry also worked intermittently at Montpelier Manufacturing through high school and college. Like Jim, Jerry was often the one who was assigned the most banal of errands in support of office or production operations. Due to a critical shortage of radiators that were required for White PDQ trucks working their way down the assembly line, Jerry was told to take a recently manufactured delivery truck and drive to the Harrison Radiator Company in Valporaiso, Indiana, 150 miles west of Montpelier. It was wintertime and the trip was very cold. The truck Jerry drove did not have a partition between the cargo and driver's compartments nor did it have a heater that was intended to provide a comfortable level of heat in northern winters. Jerry placed cardboard over the radiator in a feeble effort to coax a little more cab heat out of the heater. The truck was also equipped with a governor that limited speed to 50 MPH. An Indiana State Trooper stopped Jerry for driving too slowly on the Indiana Toll Road and ushered him to an alternate route along a state highway.

Jerry arrived at Harrison Radiator only to find that the radiators weren't ready to be picked up. He spent several hours in the delivery truck, trying his best to keep warm and get some sleep while night shift workers produced the radiators. The radiators were finally loaded into the truck at 3:00 AM and Jerry started on his way back to Montpelier. With a lower volume of interstate traffic at night, Jerry returned to the toll road. Cold and tired, he stopped at a rest area for some coffee. Backing out of his parking space, he hit a light stanchion, inflicting irreparable damage to the new truck's back bumper. Adding to his fatigue and physical discomfort was apprehension over his father's certain indignant response to the damaged bumper. Thankfully, one of the first employees Jerry encountered upon arriving at the plant was Denver Miller. As he pointed to the mangled bumper and explained the incident, he was relieved to hear Denver say, "Don't worry about it, we'll just drop it and install a new one. Your dad doesn't need to know a thing about it."

Recreation was a concept that was totally foreign to Harry Schwartz. The family had a boat at their lake property but Harry never set foot in it. Most of his evenings at home were spent in his private study where he tended to business and professional matters. Later in life he revealed to his son Jim that he shouldn't have distanced himself from his wife and children quite so much. Other than that, he was extremely content to guide the business and to experiment with new or improved delivery truck concepts as long as a steady stream of incoming revenue allowed him to do so. Jim and Jerry were sent to Howe Military Academy during their teen years out of Harry's overriding concern for providing his children with the best possible educational opportunities. To Harry, sending his boys to the academy was the essence of responsible fatherhood.

Harry paid his employees good wages, better than most area businesses. However, without the slightest hesitation, he would furlough workers each year during the October model changeover or at any other time business appeared to be tapering off, a policy that tended to offset Montpelier Manufacturing's attractive wage scale. Worker seniority was not a consideration when the time came to implement a layoff; layoffs were Harry's method of eliminating "dead wood," the marginal performers that were a drag on productivity or quality. When the model changeover was completed or business was once again steady, Harry would dispatch his foremen and superintendents at the end of a day's work to regions near and far in an effort to find better replacement workers. The recruiters would stop in at pool halls, bars or lodge halls, anywhere a group of men could be approached about employment opportunities at the body works.

Harry's concern for his employees didn't stop with a fair wage. A man who was new to the area was hired without a background check and without an existing employee to vouch for him. He was in desperate financial straits and needed money immediately. Nervously, he asked Harry for an advance on his pay only minutes after he was hired. Harry could see that the man was in need and gave him the advance without any hesitation. Jim Schwartz echoes the feelings of many Montpelier Manufacturing employees, recalling that "My father was a taskmaster but he loved his employees. He'd go the extra mile for them but he demanded loyalty. If you were loyal to him, he was loyal to you." It wasn't beyond Harry to hire an attorney if one of his employees needed legal assistance. On the other hand, you didn't want to cross him. Jim summed up his description of his father by referring to him as "old school."

Employees would come to the back door of the sprawling Schwartz ranch house a few miles west of Montpelier and ask to see Harry privately, most often around Christmas time. Fairly certain of the reason for the visit, Harry would comply, leading them back to his home office. After a momentary hesitation, the employee would confess that, on one or more occasions during the year, he had stolen tools or other articles from the factory and that he had come to tell him about it and make restitution. Harry would thank the employee for his honest admission and tell him to take what he owed for the stolen items and give it to his church as an offering. Harry sometimes told the family about the incident, intent on using it as an object lesson for his children while being very careful not to reveal the name of the individual or details of the transgression.

Fringe benefits at Montpelier Manufacturing were few and fleeting but all employees could count on receiving a frozen turkey at Christmas time. With J. D. Hill's passing in 1939, Harry's wife, Margaret was elected director and vice president of the company. Although she faithfully attended board meetings, her role in the company was entirely passive with one notable exception. After several years of distributing frozen turkeys, Margaret discovered that Harry was consistently doing so on Christmas Eve, as workers were leaving the factory. Stupefied by her husband's thoughtlessness, Margaret asked Harry how he

expected his employees to thaw their turkeys in sufficient time for a festive Christmas Day repast. Harry replied that he was unaware that thawing a frozen turkey was a prerequisite to roasting it. After Margaret enlightened him, he began distributing turkeys three days prior to Christmas.

Chuck Brigle, who held Harry Schwartz in high esteem and learned much from him as he moved from the shop floor into a drafting position and then into a manufacturing support role, recalls that, "I had a fishing place up in Michigan that I always liked to go to on Labor Day. I'd go up on Friday and come home on Tuesday. Well, there was something going on at the plant that we absolutely had to get done. I went in and wanted to know if I could have Saturday off because I wanted to go up north fishing. Asking Harry Schwartz for permission to go fishing was about the same as asking your wife or mother." Chuck had barely finished his entreaty when Schwartz shot up out of his chair. Leaning over Chuck who was seated in a leather chair across from Schwartz's desk, Schwartz replied, "'You mean to tell me that you want to go fishing while the rest of us work?" When Chuck answered in the affirmative Schwartz launched into a ten-minute dissertation on the reasons why his petition was unjustified. Chuck let him proceed through the entire speech without interruption. Harry finished by saying, "Now, if you feel like you have to have Saturday *and* Monday off, you go ahead and go. The rest of us will carry you." Refusing to buy the guilt trip on which Harry attempted to send him, Chuck answered, "I'll see you on Tuesday." That Friday evening as Chuck was about to leave work, Harry, without any explanation, handed him an envelope. With no idea of what it contained, Chuck opened it to find one hundred dollars in cash. Awestruck, Chuck collected his thoughts and hesitated a second or two before establishing eye contact with his boss. When he finally did look up, Harry cracked a smile and said, "Have a good time, Chuck."

Chuck impulsively complained one day to Harry about mounting production problems that were a drag on meeting delivery deadlines for certain trucks. In the midst of Chuck's lamentations, Harry sprang out from behind his desk and blurted, "'Oh, fine! Here, I'll give you my handkerchief. You take a corner and cry and I'll take a corner and cry. Now what does that solve? Stop whining and go get busy!" Chuck accepted the disciplinary rebuke from the "handkerchief incident" as a lifelong invitation to tackle problems instead of pointing fingers. Reflecting on this and other lessons learned from the fourteen years he worked for Harry Schwartz, Chuck summarized his thoughts by saying, "I've thought many times during my life of nearly eighty-five years about the things I learned from Harry Schwartz. After my time at the body works, I moved on and went into some pretty big companies and more than once I can recall things that he told me that helped me out.

"Hardly anyone had a good word for Mr. Schwartz and I never could understand that. The general public would call him a skinflint, forcing employees to work for nothing. It always hurt me. It just wasn't right because they really didn't know the man. They were only seeing it from their little pedestal and they weren't looking at the whole picture. He was investing a lot of money to make the place better and I certainly wouldn't have anything but praise for a man like that. He had a tough job trying to satisfy townspeople, workers, and all of his customers. If anybody needed bawled out, they got it. He didn't just pick on a few people. You don't have to be around him but a short time until you can see what the man is made of and he didn't have time to mess with a lot of mediocre stuff that slows you down."

George Cornell spent most of his time in the sheet metal department serving as a foreman under Ferman Haase. Harry admired George's loyalty and dedication to his work. When George's father passed away, Harry realized that the extended Cornell family would struggle financially with George as its sole provider and, without hesitation, gave George a substantial raise of fifty dollars a month. When it became obvious to Harry that George's right leg was bothering him from what was possibly polio, Harry paid for a thorough examination of George's leg at Parkview Hospital in Fort Wayne.

For many years, Ferman Haase was superintendent over the pattern shop, sheet metal fabrication and welding departments. Bill Sprankell was Ferman's counterpart in the priming, painting and final assembly departments. Schwartz had the diabolical and nasty habit of continually pitting Haase and Sprankell against each other. Chuck Brigle recalls that, "You could see it as plain as day. Harry even told me that he couldn't make production work very well until he got one man to scrap with the other and, between the two of them, they made production *(deadlines)*. They were in competition with each other to make sure that they weren't at fault for production delays."

Jim Schwartz received a degree in production management from Michigan State University. His education opened his eyes to more advanced manufacturing methods and he felt compelled to tell his father that "There are newer and better methods available to us. Why don't we implement them?" Harry's somewhat chilling response was, "Jimmy, don't you think that I've made a success out of this?" Jim feels that his ideas were spurned because he was young and inexperienced, even though he was a recent college graduate with knowledge that was decidedly relevant to Montpelier Manufacturing's operations. He attributes his father's disregard for his input to the fact that Jim and his brother Jerry were, and may forever be, "just boys." It's fair to look back and say that Harry sometimes handicapped himself with a malady known as the NIH (Not Invented Here) syndrome.

Harry Schwartz resisted putting money into bricks and mortar. Vast areas of the Montpelier Manufacturing shop simply were not geared for efficient production. But Harry rationalized that, given the existing condition of the plant, he could get a satisfactory return without additional investment. Coming out of the Great Depression left Harry and many others with an ingrained "bunker mentality" and a propensity for investing only on the bare essentials in operating a business. The plant badly needed a sprinkler system to suppress fires yet, in spite of his dread of fire Harry refused to install one. "In the long run, I think it was to his detriment," says his son Jim, "He always dreaded fire."

At the age of 62 Harry declared, "Enough is enough" and sold the company to White. He stayed on as President of the Montpelier Division, signing a three-year contract with White. He remained at the helm of the Montpelier Division for two years before retiring and serving the final year of his contract as a consultant after moving to Arizona. After remarrying in 1962, Harry and his new wife Hilda returned to Montpelier, Ohio and the sprawling Schwartz ranch home a few miles west of the village. Unfortunately, Montpelier and the surrounding area seemed somewhat alien to Harry and the new Mrs. Schwartz and their stay was brief. Harry lived the rest of his years comfortably in Arizona where he passed away in 1982 at the age of 86.

In the community at large, Harry had served as president of Rotary, served on the Board of Education and joined with other leaders in establishing the Williams County General Hospital. Harry was chosen to represent the community of Montpelier and speak on their behalf at a hospital dedication and did so while his wife and children were at the lake. He would typically do things, particularly if they were of a unique or special nature, without telling his immediate family. The Schwartz children were largely excluded from all of the hoopla leading up to Montpelier Manufacturing's momentous E-Award celebration because, according to Harry, it was his business, not theirs. Jim Schwartz emphasized that, "He didn't brag on himself, like his father-in-law. He was a very private individual."

Jerry Schwartz summed up his thoughts about his father by stating, "My dad was a very honorable and loving man but he was also shy and somewhat aloof, traits that undeservedly tarnished his reputation among the many people that were aware only of his prominence and ambition and not his personal character. Although he was a brilliant man, he never wore his great intellect as a self-congratulatory badge of superiority."

Jim even feels that the lack of a broad-based middle management team might have been one of the factors that prevented the company from starting over from scratch again after the fire. It wasn't so much that Harry didn't trust anyone with responsibility...it was simply that the entire operation was his and, as such, he felt that the full responsibility of managing the business should rightfully belong to him and no one else.

Bill Sprankell

Bill Sprankell was the oldest of three boys raised on a farm near Bellefontaine, Ohio. His father abandoned the family, prompting Bill to quit school after the eighth grade in order to provide as best he could for his mother and two younger brothers. Bill accepted odd jobs whenever they became available, chopping wood, working as a farm hand or delivering groceries.

He began his first full-time job in 1916, working for the Bellefontaine Division of the A. J. Miller Company, building ambulance and hearse bodies. In 1918 he moved to Grand Rapids, Michigan where he was employed at a furniture factory. He returned to work for the A. J. Miller Company in 1923, dividing his time between the Defiance and Bellefontaine, Ohio divisions. The Defiance plant was the former Turnbull Wagon factory, purchased by A. J. Miller when Turnbull operations moved to Montpelier in 1921. The A. J. Miller Company also purchased Turnbull's Montpelier plant after the Turnbull Wagon Company bankruptcy in 1925, intending to temporarily resume wagon manufacturing operations while converting the plant into its third automobile and truck body production facility, a plan which never materialized. However, when the Montpelier Manufacturing Company was organized in 1927, Bill joined the new company, serving first as assistant plant superintendent and, a year and a half later, as the plant's full superintendent. As production volumes increased in the late 1930's, plant superintendent duties were divided between Bill and Ferman Haase. Bill supervised the mounting of truck bodies to their chassis, priming, painting and undercoating, wiring, final parts mounting, and finished vehicle inspection as well as the company stockroom. Along with Harry Schwartz and Ferman Haase, Bill witnessed the growth in employment of the company from eight employees to as many as 270.

In the late '30's, Bill found it necessary to take a leave of absence due to mounting job pressures. He spent some time in Tucson, Arizona and, upon returning, sought periodic counseling from a doctor who so admired Bill that he became one of his closest and most loyal friends. Bill felt that his toughest assignment was that of gearing up for production of troop and cargo transport bodies that the company supplied to the Army during and after World War II. Bill's last years at Montpelier Manufacturing were spent directing all facets of quality control, from inspecting raw materials to personally testing each fully assembled truck prior to shipment. Confronted throughout his career with production deadlines, parts shortages, and numerous manufacturing and employee relations issues, he made it a practice to shun heated arguments by heading for home when he felt tempers were needlessly out of control. Usually by the next morning, tempers had subsided and issues could be reasonably resolved. If Bill did vent his anger at anyone, he quickly mended affairs, never allowing the sun to set before his wrath had dissipated.

Next to his faithful commitment as a long time member of St. Paul's Methodist Church in Montpelier, Bill was an intensely devoted husband and father. Without fail he would help his wife, Jessie, with the dishes after the evening meal, never missing an opportunity to give her a kiss as he finished the kitchen chores. He was an avid gardener and it was his longtime tradition to plant pink geraniums each year in three planters below the marquee of St. Paul's Church. After retiring, Bill could be seen taking his daily walks in downtown Montpelier, waving cordially to his many friends and admirers. He was a widower for eight years prior to his passing in 1986 at the age of 86.

Ferman Haase

Sources vary as to when Ferman Haase joined Montpelier Manufacturing but his work background before joining the firm was certainly varied and interesting. Born in 1894 in Evansport, Ohio, Ferman's formal schooling ended after the sixth grade when he took a job as a carpenter's assistant, building barns. He and his new bride moved to Toledo in 1916 where he worked first for Willys-Overland and, during World War I, at the Toledo Shipyards where he hung doors on hospital ships. After the war, he partnered with his brother and brother-in-law in purchasing a lumberyard. Ferman and his brother sold their interests in the lumberyard and became successful home builders in Fort Wayne, Indiana. The stock market crash of late 1929 brought their contracting business to an abrupt end and the brothers were forced to declare bankruptcy and forfeit 32 homes that they had built on speculation during the bustling economic period immediately before the crash.

Jobs were difficult to find as the 1920s drew to a close, but a friend of Ferman's directed him to Harry Schwartz and a job at Montpelier Manufacturing. Drawing upon his skills as a builder, Ferman was hired as a pattern maker and, by 1939, was elevated to the position of plant co-superintendent, responsible for supervising the body framing area, body panel fabrication, and welding.

Ferman joined Ken Harman and Chuck Brigle in the engineering department after a company reorganization in 1943. His innate abilities as a manufacturing and tooling engineer were responsible for bringing new or modified truck body designs to fruition and for steadily improving the company's manufacturing processes. He designed and developed tooling that rapidly folded metal door skins over their plywood cores and devised improved methods for converting the conventional steering arrangement on incoming truck chassis to forward control steering.

Ferman suffered the partial loss of three fingers while conducting proof tests on Montpelier's Navy catapult. In another test of the catapult, Ferman ill-advisedly oiled the piston and internal seals of the air cylinder that dissipated the catapult's energy after its projectile was launched. With the seals and piston lubricated, the trapped air in the cylinder was compressed so fast that enough heat was generated to ignite the lubricating oil. The resulting explosion was deafening, the cylinder was blown into shreds and the whole sequence of events was over before the stunned test crew could blink or run for cover. It required several minutes, however, before normal cardiac activity resumed for those who had just observed the phenomenon.

During the war, Ferman roamed the area surrounding Montpelier on many an evening, seeking suitable new employees for Montpelier Manufacturing. A chronic shortage of labor forced Montpelier Manufacturing employees to work long hours to meet wartime production quotas, a fact that made achieving the Army-Navy "E" award all the more remarkable. Ferman succeeded in getting his two brothers-in-law to join the company but he made sure that they were given the most distasteful and thankless jobs in the plant, suppressing any speculation that he was showing them favoritism. His in-laws did the work that was required of them but left the company after fully realizing the hardships that came as a result of their relationship to Ferman. On nights when he wasn't searching for potential employees, Ferman would come home for dinner at 5:00 and return to the factory to continue working until 10:00.

By 1946, Ferman had worked at Montpelier Manufacturing for five years without a vacation. At war's end, Harry Schwartz rewarded him with a *three month* paid vacation. Ferman was able to purchase a brand new Packard Clipper sedan, the first to arrive at the Earl Wolff Dealership after post-war passenger car production resumed. He used the lengthy vacation and new car to take the family on a tour of the southwest United States.

Ferman was a man of boundless energy, the type of employee that Harry Schwartz wished that he could clone over and over. A very high-strung individual that moved at a hectic pace from one mission to the next, Ferman's slowest gait was an aerobic workout in the eyes of everyone who knew him or worked for him. He was known for his ability to swing one leg backwards and do a back flip even past the age of sixty!

Shop workers could gage Ferman's temperament at any given moment by the orientation of his hat. When his hat was tipped back on his head, he was in good humor. Whenever the hat was pulled down over his forehead, it was certain that he had just emerged from the office after an argument with Bill Sprankell and/or Harry Schwartz and it was time to keep busy and leave him alone.

Ferman stayed with Montpelier Manufacturing until February 1962 when he retired at the age of 68. While there, he never failed to draw a weekly paycheck, although during lean times he sometimes only worked a half-day per week. After retirement he was hired by Montpelier Manufacturing as a part time consultant while helping out on his son Frank's farm. He also worked on a contract basis for a Goshen, Indiana truck body builder, developing a van body with a forward sliding roof section and a work platform that was elevated with a pantograph linkage. Ferman was constantly in search of a better way to get work done, not only at work but also at home. He had Montpelier Manufacturing employees make a fruit tree sprayer to his design specifications and built his own reel-type power lawnmower years before power mowers were commercially available. Ferman's motivation for increasing productivity at work was two-fold; he wanted to maximize overall production efficiency for the benefit of the company *and* show shop floor workers how they could be more productive in order to increase their piece rate wages. His son Frank describes him as an irrepressible motivator *and* dictator and made it a point to say, "You didn't want to cross him."

Photo Archive

A circa 1930 Mack Model BC or BJ cowl and tractor chassis equipped with a Montpelier Deluxe Model 105 sleeper cab.

A Montpelier cab-over-engine (COE) cab mated a 1937 Chevrolet chassis. Montpelier made the necessary steering and suspension modifications to a conventional Chevrolet truck chassis before their COE cab could be installed. Chevrolet offered this Montpelier cab for the 1937 model year only while Dodge made this cab available from 1937 through 1939. The manufacturer of the insulated box van body for this Churngold Margarine truck is unknown.

Montpelier's Side-Aisle and Low Step Urban trucks were on prominent display at the Detroit Zone Chevrolet Truck and Equipment Show in 1946.

A 1950 Chevrolet 1-ton cowl and chassis unit after removing over twenty inches of frame rail and installing a Montpelier fabricated "drop box" or stand-drive unit. Conventional driver controls for operation while the driver is seated are visible as is a combination brake and clutch pedal that was operated by the driver while standing.

The Jewel Tea Company was a long-standing customer of Montpelier Manufacturing and operated as many as 12,000 trucks at one time, offering home delivery of gourmet foods and fine quality household products Most forward control delivery trucks were equipped with dual rear doors but this model has the less costly single door with an optional extra cost window.

A good looking delivery truck and a neatly attired driver put the finishing touches on a customer purchase, leaving a final impression of the transaction that Sears and other retail stores hoped would lead to repeat business. This Sears Urban briefly plied the streets of Montpelier before settling into its daily delivery routine in a larger metropolis.

Two-tone paint treatment was an extra cost option on Montpelier Urban Delivery trucks but many retail stores paid the premium in order to have a truck that created a good visual impression with customers on a multi-stop delivery route. This Model 8 Urban Delivery mounted on a Chevrolet 3/4-ton or 1-ton chassis for the Grand Union Tea Company would look just as appealing on a delivery route today as it did when new in 1951.

The Bowman Dairy Company operated this shortened 1955 Dodge 1-ton cowl and chassis with its Montpelier Side-Aisle body for home delivery of its dairy products in Northwest Ohio. Features worth noting on this truck include the modified "drop-box" truck frame incorporating the Side-Aisle, a sliding door partition between the driver and cargo compartments and a fairly elaborate grille guard.

Ford provided Montpelier Manufacturing and other body builders with stripped chassis as well as chassis equipped with a front cowl and windshield. The sloping roofline over the driver's compartment of this truck reveals that the truck is equipped with a Montpelier body mated to the Ford-supplied front clip. The vast majority of Montpelier-bodied forward control delivery trucks were built on stripped chassis, making this truck a rare exception.

A 1959 Montpelier Low-Step Urban Body mounted on a 1959 Chevrolet forward control chassis near Montpelier, Ohio. The salvage yard owner has tentative plans to one day turn this vehicleinto an emergency road service vehicle. Somebody with more than a tentative plan needs to rescue this truck. Fiberglass roof appears to be in good shape, flat glass and rubber windows seals would be relatively easy to replace. Your author's cost in acquiring this photo: one junkyard dog bite.

Montpelier filled a large order for Side-Aisle Delivery truck bodies mated to Ford chassis in 1961. A small portion of the order is shown here. The trucks have a welded wire front-hinged side door in addition to the conventional sliding side door.

White Motor Company provided PDQ demonstrator models to dealers in areas where sales potential was significant. The company also put together a thorough and well-conceived sales program for its dealers. White dealers lamented that the demonstrator models weren't aimed at the heart of the delivery truck market. In reality, White dealers either failed to aggressively compete for sales or found themselves unable to compete with the more savvy International truck dealers or the new and much less costly Ford Econoline compact delivery vans.

A power and front suspension module equipped with a Perkins 4-236 diesel. The four-cylinder Perkins powerplant eventually became an option to gasoline fueled engines for all White PDQ delivery trucks. The radiator is equipped with shutters and a "shutter stat", a thermostat device that initiates the opening of the shutters only after radiator cooling is required.

Continental Baking Company was the Montpelier Division's first customer for these modern Urban Delivery trucks and played a significant role in their design. This transcendent design looks just as modern today as it did when first introduced in 1964. Wonder Bread is now a product of the Interstate Brands Corporation but their white forward control delivery trucks with red wheels are still a familiar site along wholesale bread delivery routes.

This state-of-the-art aluminum and fiberglass reinforced plastic Urban Delivery body was assembled to a Chevrolet light-duty truck chassis and made a brief circuit through the finer neighborhoods of Montpelier for a photo session before it departed with many similar trucks for daily UPS local delivery service.

Montpelier's first business deal with Nickles Bakery of Navarre, Ohio involved equipping their sedan delivery trucks with bakery racks. Once Nickles management familiarized themselves with Montpelier's fanatical attention to customer needs, they began ordering Montpelier Side-Aisle Delivery trucks in the mid-1950's. By 1964, they had moved up to Montpelier's latest and greatest offering, the high-top Urban Delivery truck of aluminum and fiberglass construction.

WE DROVE THE WHITE HORSE *Over a Cliff!*